Privacy-Preserving Data Publishing

An Overview

Synthesis Lectures on Data Management

Editor
M. Tamer Özsu, *University of Waterloo*

Synthesis Lectures on Data Management is edited by Tamer Özsu of the University of Waterloo. The series will publish 50- to 125 page publications on topics pertaining to data management. The scope will largely follow the purview of premier information and computer science conferences, such as ACM SIGMOD, VLDB, ICDE, PODS, ICDT, and ACM KDD. Potential topics include, but not are limited to: query languages, database system architectures, transaction management, data warehousing, XML and databases, data stream systems, wide scale data distribution, multimedia data management, data mining, and related subjects.

Privacy-Preserving Data Publishing: An Overview
Raymond Chi-Wing Wong and Ada Wai-Chee Fu
2010

An Introduction to Duplicate Detection
Felix Naumann and Melanie Herschel
2010

Keyword Search in Databases
Jeffrey Xu Yu, Lu Qin, and Lijun Chang
2010

Copyright © 2010 by Morgan & Claypool

All rights reserved. No part of this publication may be reproduced, stored in a retrieval system, or transmitted in any form or by any means—electronic, mechanical, photocopy, recording, or any other except for brief quotations in printed reviews, without the prior permission of the publisher.

Privacy-Preserving Data Publishing: An Overview
Raymond Chi-Wing Wong and Ada Wai-Chee Fu
www.morganclaypool.com

ISBN: 9781608452163   paperback
ISBN: 9781608452170   ebook

DOI 10.2200/S00237ED1V01Y201003DTM002

A Publication in the Morgan & Claypool Publishers series
SYNTHESIS LECTURES ON DATA MANAGEMENT

Lecture #3
Series Editor: M. Tamer Özsu, *University of Waterloo*
Series ISSN
Synthesis Lectures on Data Management
ISSN pending.

Privacy-Preserving Data Publishing

An Overview

Raymond Chi-Wing Wong
The Hong Kong University of Science and Technology

Ada Wai-Chee Fu
The Chinese University of Hong Kong

SYNTHESIS LECTURES ON DATA MANAGEMENT #3

 MORGAN & CLAYPOOL PUBLISHERS

ABSTRACT

Privacy preservation has become a major issue in many data analysis applications. When a data set is released to other parties for data analysis, privacy-preserving techniques are often required to reduce the possibility of identifying sensitive information about individuals. For example, in medical data, sensitive information can be the fact that a particular patient suffers from HIV. In spatial data, sensitive information can be a specific location of an individual. In web surfing data, the information that a user browses certain websites may be considered sensitive. Consider a dataset containing some sensitive information is to be released to the public. In order to protect sensitive information, the simplest solution is not to disclose the information. However, this would be an overkill since it will hinder the process of data analysis over the data from which we can find interesting patterns. Moreover, in some applications, the data must be disclosed under the government regulations. Alternatively, the data owner can first modify the data such that the modified data can guarantee privacy and, at the same time, the modified data retains sufficient utility and can be released to other parties safely. This process is usually called as privacy-preserving data publishing. In this monograph, we study how the data owner can modify the data and how the modified data can preserve privacy and protect sensitive information.

KEYWORDS

privacy preservation, data publishing, anonymity, data mining

Contents

1 Introduction 1
    1.1 Data Publishing 1
    1.2 Significance 3
    1.3 Organization 4

2 Fundamental Concepts 7
    2.1 Anonymization 9
    2.2 Information Loss Metric 18
    2.3 Privacy Models 19
    2.4 Other Privacy Models 21
    2.5 Conclusion 28

3 One-Time Data Publishing 29
    3.1 Knowledge about Quasi-identifiers 29
    3.2 Knowledge about the Distribution of Sensitive Values ... 40
    3.3 Knowledge about the Linkage of Individuals to Sensitive Values ... 44
        3.3.1 Information That Some Individuals Do Not Have Some Sensitive Values   44
        3.3.2 Information That Some Individuals Have Some Sensitive Values   45
    3.4 Knowledge about the Relationship among Individuals 46
    3.5 Knowledge about Anonymization 47
    3.6 Knowledge Mined from the Microdata 52
    3.7 Knowledge Mined from the Published Data 58
    3.8 How To Use Published Data 60
        3.8.1 Aggregate Queries   61
        3.8.2 Information Loss   62

		3.8.3 Evaluation with Data Mining and Data Analysis Tools 64

        3.8.4 Querying over an Uncertain Database 65

  3.9 Conclusion 66

4 Multiple-Time Data Publishing 69

  4.1 Individual-Based Correlation 69

        4.1.1 Data publishing from Static Microdata 69

        4.1.2 Data Publishing from Dynamic Microdata 73

  4.2 Sensitive Value-Based Correlation 80

        4.2.1 Protection for Permanent Sensitive Values 83

        4.2.2 Protection for Transient Sensitive Values 85

  4.3 Conclusion 87

5 Graph Data 89

  5.1 Data Model 89

  5.2 Adversary Attacks 90

        5.2.1 Assumption of Adversary Knowledge 90

        5.2.2 Active Attacks 92

  5.3 Utility of the Published Data 93

  5.4 k-Anonymity 94

        5.4.1 Vertex Degree 94

        5.4.2 1-Neighborhood 95

        5.4.3 Vertex Partitioning 96

        5.4.4 k-Automorphism 98

  5.5 Multiple Releases of Data Graphs 100

  5.6 Other Approaches 101

  5.7 Future Directions 101

6 Other Data Types 103

  6.1 Spatial Data 103

        6.1.1 With Anonymizer 104

        6.1.2 Without Anonymizer 107

	6.2	Transactional Data 108
	6.3	Conclusion 111

7 Future Research Directions 113

	7.1	One-Time Data Publishing 113
	7.2	Multiple-Time Data Publishing 114
	7.3	Publishing Graph Data 114
	7.4	Publishing Data of Other Forms 115

A Definition of Entropy l-Diversity and Recursive l-Diversity 117

Authors' Biographies 127

CHAPTER 1

Introduction

Privacy preservation has become a major issue in many data mining applications. When a data set is released to other parties for data mining, privacy-preserving techniques are often required to reduce the possibility of identifying sensitive information about individuals.

1.1 DATA PUBLISHING

Consider a party A that releases its own data T to another party B for data mining. Party A is the data owner and party B is the data miner. In a more general setting, B corresponds to the public, which means that party A releases T to the public. We call this process *data publishing*.

If T contain no sensitive information, A can directly pass T to B. However, in most cases, T contain some sensitive information and thus A cannot give T to B in its raw format.

Sensitive information appears in many applications. For example, in medical data, sensitive information can be the fact that a particular patient suffers from HIV. In spatial data, sensitive information can be the specific location of an individual. In web surfing data, the information that a user browses certain websites may be considered sensitive.

In order to protect sensitive information, one naive solution is *not* to disclose the information. However, this is an overkill since it will hinder mining the data that can reveal interesting patterns. Besides, in some applications, the data must be disclosed under government regulations. For example, the hospitals in California have to disclose patient records on the Web [Carlisle et al., 2007].

Alternatively, the data owner first *modifies* the data (T) such that the modified data (T^*) contain no sensitive information so that the modified data can be released to other parties safely. This process is usually named *privacy-preserving data publishing*.

The major challenge of data publishing is to modify the data such that the modified data contain no sensitive information. This raises three questions:

- How does the data owner modify the data?
- How does the data owner guarantee that the modified data contain no sensitive information?
- How much does the data need to be modified so that no sensitive information remains?

Let us first give brief answers for these questions. We will answer them in more detail in the remainder of this book.

For the first question, there are many techniques to modify the data. One way is *perturbation*, under which the data content is changed (randomly). For example, if the medical data contain a record showing that a patient lives in San Francisco, data perturbation may change the address of this patient to Boston. Another technique is *generalization*, that changes the data content to less specific information. For instance, under generalization, the address of a patient would be changed from San Francisco to California.

Consider the second question. If the data owner makes sure that an *adversary* or an *attacker* cannot infer any sensitive information from the modified data, then the modified data contain no sensitive information. Figure 1.1 shows an adversary who tries to infer sensitive information from the modified data. Usually, the adversary is equipped with some *background knowledge* that can be used as additional information to infer sensitive information from the modified data. There are many types of background knowledge. One is the *presence* knowledge. Consider that a hospital releases its medical data containing all patient medical records where each record contains the district and the disease of a patient. If the adversary knows that a neighbor who lives in the same district has visited this hospital and there is only one record with this district in the released data, the adversary can deduce that this record corresponds to this neighbor. If the disease attribute of this record has value HIV, then s/he can further infer that the neighbor suffers from HIV. There are many other kinds of background knowledge, and we discuss these in Chapters 2 and 3.

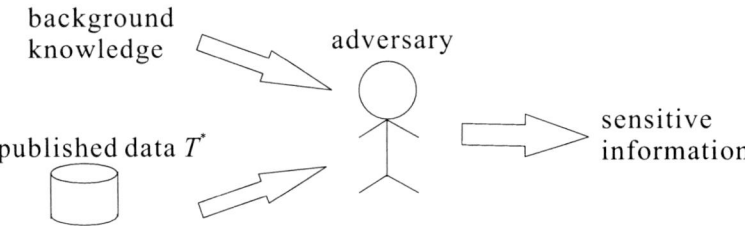

Figure 1.1: An adversary who tries to infer sensitive information from the modified data

Consider the third question. In order to generate data without sensitive information, the data owner has to modify the data. However, when we modify the data, we distort the usefulness of the data. There is a tradeoff between privacy and usefulness. There are *many* ways to modify data such the modified data contain no sensitive information (Figure 1.2). Which modified data should the data owner select for publishing? The best modified dataset for data publishing is the one with the highest *utility*. Generally, if the modified data have high utility, then the data contain more useful information for data analysis. For example, consider that we adopt the generalization technique for data modification. The change from San Francisco to California introduces less information loss than the change from San Francisco to United States. Thus, intuitively, the modified data that contain California have higher utility than the modified data that contain United States. We will give a detailed description of utility in Chapter 2 and Chapter 3.

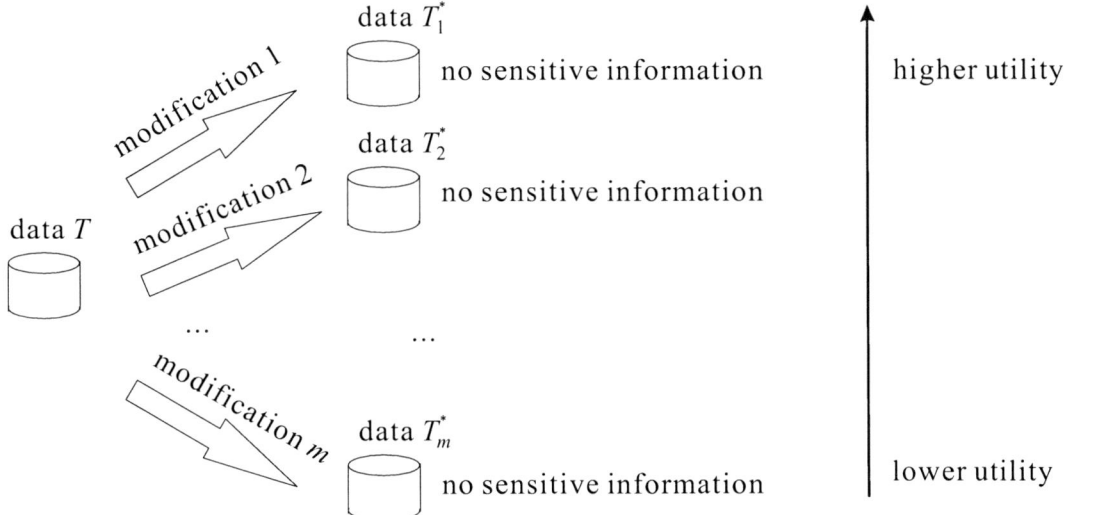

Figure 1.2: Multiple modified data that contain no sensitive information

1.2 SIGNIFICANCE

Privacy-preserving data publishing is very important in our daily lives because there can be serious consequences if the data owner releases data without considering the protection of sensitive information. Some real events that demonstrates the problem are the following:

- In 2002, Sweeney [Sweeney, 2002b] identified the insufficient protection of a medical dataset. In a real medical data, about 87% of individuals can be uniquely identified by matching certain attributes with a publicly available external table such as a voter registration list by a simple mapping operation (details will be described in Chapter 2).

- In 2006, AOL did not take sufficient precautions and encountered some undesirable consequences. A dataset including search logs was published in 2006. Later, AOL realized that a *single* 62 year old woman living in Lilburn (in Georgia) can be identified from the search logs by New York Times reporters using her several individual-specific queries (e.g., finding webpages with her last name and finding landscapers in Lilburn (in Georgia)). The search logs were withdrawn and two employees responsible for releasing the search logs were fired [Barbaro and Jr, 2006].

- Netflix is a popular online movie rental service with a recommender system, called Cinematch, that recommends movies to its customers based on their predicted movie preferences. Netfix released its data to the public on October 2, 2006 for a challenging competition called the Netflix Prize with the aim of improving the prediction accuracy of the recommender system. However, Narayanan and Shmatikov [2006] found that 96% of the subscribers could be

4 1. INTRODUCTION

uniquely identified by limited knowledge of at most 8 movie ratings with their corresponding rating dates.

- The research project conducted by Beinat [2001] shows that 24% of the mobile clients using location-based services (LBS) have serious privacy concerns about disclosing their locations together with their personal information.

1.3 ORGANIZATION

We organize this book according to *data form* and the *frequency of data publishing*.

- *Data form:* There are different forms of data to be released. A relational table is a common form of data. Spatial data, transactional data and graph data are some other forms of data. In this book, first, we focus on relational table to illustrate the concept of data publishing. Then, we discuss how data publishing can be applied to other forms of data.

- *Frequency of data publishing:* Earlier, we considered cases where the data owner publishes the data only *once*. However, typically, the data owner publishes the data *multiple* times at different time instances, as shown in Figure 1.3, which is referred to as *multiple-time data publishing* or *sequential data publishing*. Multiple-time data publishing is more challenging compared to one-time data publishing. This is because the adversary can infer sensitive information by using the knowledge about the correlations among all past published data.

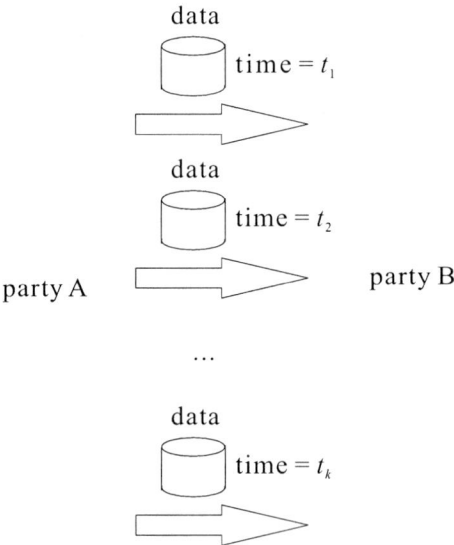

Figure 1.3: Multiple-time data publishing

The rest of the book is organized as follows. Chapter 2 describes fundamental concepts of data publishing. Chapter 3 discusses details of one-time data publishing based on the fundamental concepts developed in Chapter 2. Chapter 4 considers how to perform multiple-time data publishing. Chapter 5 discusses how to anonymize graph data. Chapter 6 extends the discussion to handling data publishing over data of other forms. Chapter 7 gives some directions for future work.

CHAPTER 2
Fundamental Concepts

In this chapter, we discuss some fundamental concepts used in anonymizing data. In order to better describe the concepts, we illustrate them using the *relational model* of data.

We consider a medical application for illustration. Consider a table that stores patient information with five attributes:

- Name (or Identifier)
- Gender
- Nationality
- Age
- Disease

The first attribute, Name, corresponds to the name of a patient. The next three attributes, namely Gender, Nationality and Age, are some personal information of the patient. The last attribute, Disease, records the illness that was diagnosed in the hospital. We call this attribute a *sensitive* attribute because it contains *sensitive values* like HIV and Cancer. Although Disease may also contain sensitive values, it may contain *non-sensitive values* like flu. A tuple associated with a sensitive value is called a *sensitive tuple*. Table 2.1 shows an example of this medical data. Assume that each tuple in the table refers to a patient, and that each patient has at most one tuple.

This table is called a *micro-data*, referring to raw data that have not been modified.

Table 2.1: A medical table

Name	Gender	Nationality	Age	Disease
Peter	male	Japanese	26	HIV
John	male	Malaysian	30	flu
Mary	female	American	36	HIV
Sally	female	Canadian	40	HIV
Eason	male	American	40	flu
Louis	male	Chinese	36	flu

In this example, the information about any particular individual (e.g., Peter) suffering from a sensitive disease, HIV, is considered sensitive. One of the objectives of privacy preserving data publishing is to hide this kind of information when data are published.

In Table 2.1, since Name can disclose the identity of a patient, the data owner removes Name from Table 2.1 and releases it as shown in Table 2.2.

Table 2.2: A published table when the adversary has no background knowledge

Gender	Nationality	Age	Disease
male	Japanese	26	HIV
male	Malaysian	30	flu
female	American	36	HIV
female	Canadian	40	HIV
male	American	40	flu
male	Chinese	36	flu

Under the assumption that the adversary has no background knowledge, Table 2.2 does not disclose any information that a particular individual suffers from a certain disease. For example, from Table 2.2, although the adversary knows that the first tuple suffers from HIV, s/he does not know that this tuple corresponds to Peter, and thus cannot infer that Peter suffers from HIV.

In reality, the adversary typically has some background knowledge that can be obtained easily from other channels. In this case, the privacy protection provided in Table 2.2 is insufficient.

In Table 2.2, the three attributes, Gender, Nationality and Age, correspond to the personal information of individuals. They are called *quasi-identifier (QI)* attributes. Quasi-identifier attributes are those that can serve as an identifier for an individual. Sweeney [2002b] points out that in a real dataset about 87% of individuals in a medical data were uniquely identified by three QI attributes, namely sex, date of birth and 5-digit zip code, using a publicly available external table such as a voter registration list. There are many sources of such a table. Most municipalities sell population registers that include the identifiers of individuals along with basic demographics; examples include local census data, voter lists, city directories, and information from motor vehicle agencies, tax assessors, and real estate agencies [Samarati and Sweeney, 1998]. Sweeney [2002b] also reports that a city's voter list in two diskettes was purchased for twenty dollars, and it was used to identify medical records. Table 2.3 shows an example of a voter registration list that contains individuals' names and the quasi-identifier attributes. Note that the total number of tuples in this table is larger than the total number of tuples in Table 2.2.

Consider that the adversary obtains the published data of Table 2.2 and the voter registration list of Table 2.3. S/he can deduce that the first tuple in Table 2.2 corresponds to Peter. Thus, s/he infers that Peter suffered from HIV. Consequently, the *individual privacy* of Peter is compromised.

In the following, for clarity, we use the terms "individual" and "tuple" interchangeably when the adversary has knowledge of the quasi-identifiers of individuals.

Table 2.3: A voter registration list

Name	Gender	Nationality	Age
Peter	male	Japanese	26
John	male	Malaysian	30
Mary	female	American	36
Sally	female	Canadian	40
Eason	male	American	40
Louis	male	Chinese	36
Helen	female	Chinese	80
Iris	male	American	40

2.1 ANONYMIZATION

Since the data owner only removes attribute Name from the original table (Table 2.1), the resulting released table (Table 2.2) discloses sensitive information.

In order to protect sensitive information, the data owner has to *anonymize* the table such that the resulting table protects sensitive information. There are different ways of *anonymizing* the table before it is published. We describe two methods, namely *grouping-and-breaking* and *perturbation*.

- *Grouping-and-Breaking:* This is an operation that divides the records horizontally into a number of partitions and then breaks the exact linkage between the QI value and the sensitive value in each partition [Sweeney, 2002a; Xiao and Tao, 2006b]. Figure 2.1 illustrates the grouping-and-breaking operation over Table 2.2.

  The goal of grouping is to make individuals in the same group indistinguishable so that the adversary cannot uniquely identify each individual in the group. The objective of breaking is to weaken the linkage between the QI values and the sensitive values so that the adversary has less confidence in the linkage that can be inferred between an individual (which can be identified by the QI values) and a sensitive value in the sensitive attribute such as HIV. How to implement the grouping step and the breaking step will be described with the following three representative concrete concepts, namely *suppression*, *generalization* and *bucketization*.

  - *Suppression:* Suppression is an operation that changes a specific value in an attribute to *ANY*, denoted by *. Table 2.4 is a table generated by suppression.

    We can see that the first two tuples and the last two tuples have the same QI values. These four tuples are said to form an *equivalence class*. An *equivalence class* of a table is a collection of all tuples in the table containing identical QI values. The third and the fourth tuples form another equivalence class. An equivalence class is also called a *QI-group*.

    Note that, since each individual is indistinguishable from the others in the same QI-group, the exact linkage between the QI values and the sensitive value in this group is broken. Consider the first QI-group containing four tuples. According to the voter registration

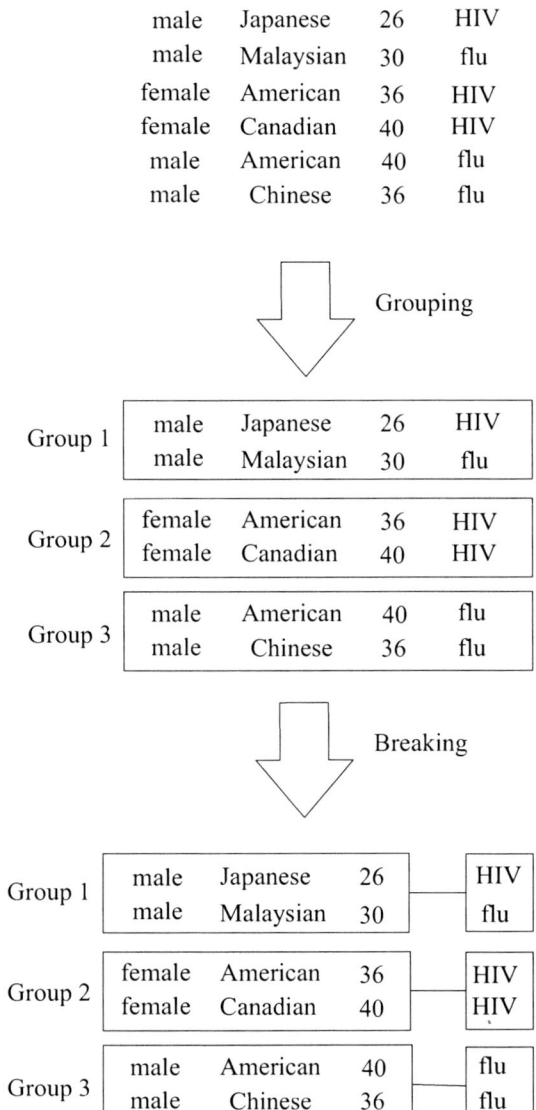

Figure 2.1: Grouping-and-breaking

list, the adversary knows that one of these four tuples corresponds to Peter, but his exact disease cannot be inferred since the whole QI-group (containing four indistinguishable individuals) has one HIV value and three flu values.

- *Generalization:* Generalization is an operation that changes a value to a more generalized one. If the value is numeric, this value may be changed to a range of values. For example,

2.1. ANONYMIZATION

Table 2.4: A published table by suppression

Gender	Nationality	Age	Disease
male	*	*	HIV
male	*	*	flu
female	*	*	HIV
female	*	*	HIV
male	*	*	flu
male	*	*	flu

value 26 can be changed to range 26-30. If the value is a categorical value, it may be changed to another categorical value denoting a broader concept of the original categorical value. For instance, city San Francisco can be changed to state California.

Usually, the generalization of a categorical attribute is based on a *generalization taxonomy* representing the semantics of the attribute. Figure 2.2(a) shows the generalization taxonomy for Nationality.

In a taxonomy, a value located at a lower level corresponds to a more specific value. In particular, the value at the lowest level corresponds to the most specific value and the value at the highest level corresponds to the least specific value. An edge from a value v_1 at a lower level to a value v_2 at a higher level corresponds to the fact that v_1 can be generalized to v_2.

The concept of the generalization taxonomy is not limited to categorical attributes. It can be extended to handling numeric attributes. In particular, we can have a hierarchical structure defined with value, interval, *, where value is the raw numeric values, interval is the range of the raw numeric values and * is a symbol representing any numeric values. Figure 2.3(a) shows the generalization taxonomies for attributes Nationality and Age, respectively.

The set of all possible values for a particular level forms a *generalization domain*. Different levels correspond to different generalization domains. For instance, in Figure 2.2(a), there are five possible values, Japanese, Malaysian, Chinese, American and Canadian, which form a domain $N0$. Similarly, Asian and North American form another domain $N1$, and Person forms domain $N2$. Figure 2.2(b) shows all possible domains where each edge from domain D_1 to domain D_2 corresponds to the fact that a value in D_1 can be generalized to a value in D_2.

Table 2.5 is an example of the data generated by generalization where Asian nationalities like Japanese and Malaysian are generalized to Asian, and North American nationalities like Canadian and American are generalized to North American.

12 2. FUNDAMENTAL CONCEPTS

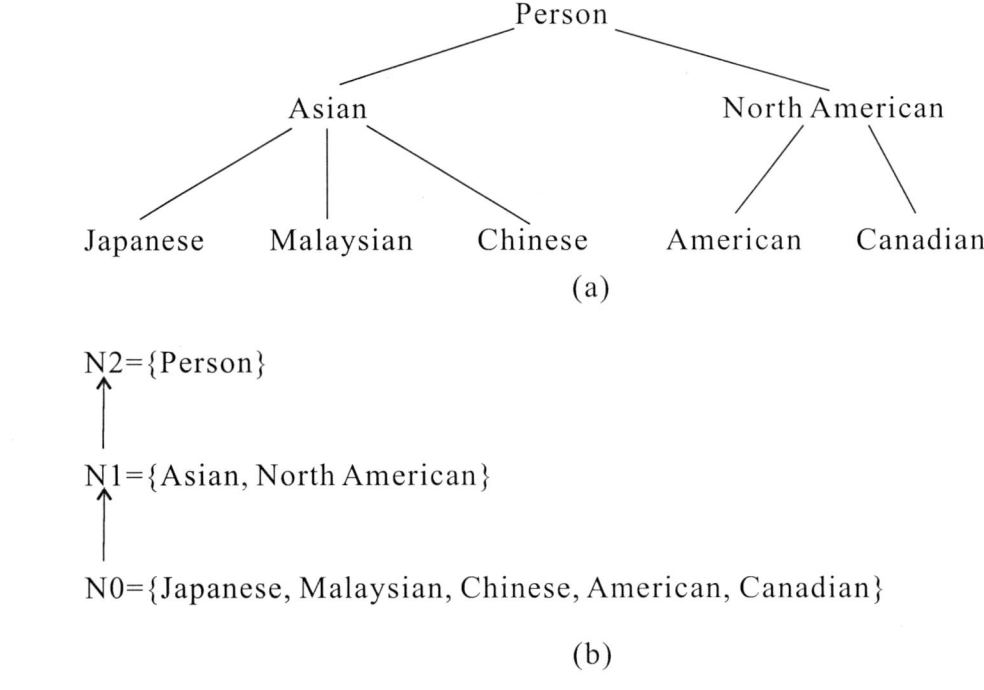

Figure 2.2: A taxonomy for attribute Nationality

Table 2.5: A published table by generalization

Gender	Nationality	Age	Disease
male	Asian	26-30	HIV
male	Asian	26-30	flu
female	North American	36-40	HIV
female	North American	36-40	HIV
male	Person	36-40	flu
male	Person	36-40	flu

Given a microtable T, a table T^* is said to be a *fully generalized table* of T if all values of each attribute in T are generalized to the value located at the root of the generalization taxonomy in T^*. Table 2.6 is a fully generalized table of Table 2.2.

Similar to suppression, it is easy to see that the first two tuples form a QI-group and the linkage between the QI values and the sensitive value is broken.

Note that suppression is a special case of generalization. This is because when we generalize a specific value to *ANY*, generalization becomes suppression.

2.1. ANONYMIZATION

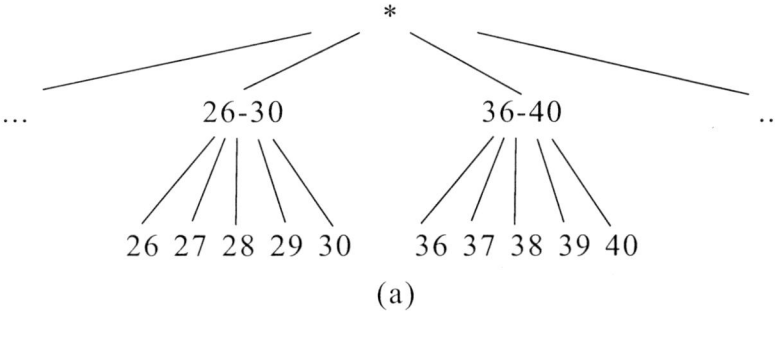

Figure 2.3: A taxonomy for attribute Age

Table 2.6: A fully generalized table of Table 2.2

Gender	Nationality	Age	Disease
*	Person	*	HIV
*	Person	*	flu
*	Person	*	HIV
*	Person	*	HIV
*	Person	*	flu
*	Person	*	flu

There are two major principles to anonymize a table using generalization, namely *global recoding* and *local recoding*.

* In *global recoding*, all values of an attribute in the published table come from the same domain in the hierarchy. For example, all values in Birth date are in years, or all are in both months and years. All values in Nationality are related to countries or continents. An entire anonymized table has uniform domains for each of its attributes, but it may lose more information. For example, a global recoding of

Table 2.2 by generalization can be Table 2.7, which suffers from *over-generalization*. Note that Table 2.5 is not a global recoding of Table 2.2 because some tuples have nationality related to continent and some have nationality related to Person.

Table 2.7: A global recoding of Table 2.2 by generalization

Gender	Nationality	Age	Disease
male	Person	26-30	HIV
male	Person	26-30	flu
female	Person	36-40	HIV
female	Person	36-40	HIV
male	Person	36-40	flu
male	Person	36-40	flu

* With *local recoding*, values may be generalized to different generalization domains. For example, Table 2.5 is a local recoding of Table 2.2 by generalization. In fact, one can say that local recoding is a more general model, and global recoding is a special case of local recoding.

  Local recoding has a good feature that the table it generates has fewer generalized values compared with the table generated by global recoding. Thus, the table generated by local recoding is more *similar* to the original table, and thus the data analysis based on this table is more accurate. However, local recoding cannot give as consistent a representation of the anonymized table as global recoding, and thus it is difficult for users to analyze the anonymized table generated by local recoding.

– *Bucketization:* Bucketization is similar to generalization, but it does not modify any QI attribute or sensitive attribute. Instead, after it divides the records into a number of partitions, it assigns a unique ID called GID to each partition, and all tuples in this partition are said to have the same GID value. Then, two tables are formed, namely the *QI table* (Table 2.8(a)) and the *sensitive table* (Table 2.8(b)), such that all tuples are projected on all QI attributes and GID to form the QI table, and on the sensitive attribute (Disease) and attribute GID to form the sensitive table.

Note that the grouping formed by Table 2.8 is equivalent to the grouping formed by Table 2.5, except that Table 2.8 contains all the original tuple values while Table 2.5 contains some generalized tuples values.

Generalization has the advantage of providing a representation with consistent attribute values within each group, which makes the analysis of the published data easier. Due to the inconsistent attribute values within each group, bucketization requires some sophisticated analysis of the data generated by bucketization. We discuss this in detail in Section 3.8.

However, since generalization changes some specific values to generalized values, the data analysis is different since the specific values are lost. Bucketization has the advantage of allowing users to obtain the original specific values for data analysis.

Table 2.8: A published table by bucketization

Gender	Nationality	Age	GID
male	Japanese	26	1
male	Malaysian	30	1
female	American	36	2
female	Canadian	40	2
male	American	40	3
male	Chinese	36	3

(a) QI table

GID	Disease
1	HIV
1	flu
2	HIV
2	HIV
3	flu
3	flu

(b) Sensitive table

- *Perturbation:* Under perturbation, a value can be changed to any arbitrary value. For example, Japanese can be changed to Chinese. Table 2.9 shows an example with perturbation where the value of Nationality of the first tuple is changed from Japanese to Chinese.

Table 2.9: A published table by perturbation

Gender	Nationality	Age	Disease
male	Chinese	26	HIV
male	Malaysian	30	flu
female	American	36	HIV
female	Canadian	40	HIV
male	American	40	flu
male	Chinese	36	flu

The following three methods, namely adding noise, value swapping and model-fitting-and-regenerating, show concrete implementations of perturbation.

- *Adding Noise:* Adding noise is a technique applicable to numeric attributes. If the original numeric value is v, adding noise will change the value to $v + \triangle$ by adding a value $\triangle$ that follows some distribution. Table 2.10 shows the anonymized table by adding noise to attribute Age where $\triangle$ comes from a normal distribution with mean 0 and standard derivation 5.

  The advantage of adding noise is that it maintains some statistical information such as means and correlations [Kim and Winkler, 1995]. However, adding noise may introduce some values that do not exist in the real world.

Table 2.10: A published table by adding noise

Gender	Nationality	Age	Disease
male	Japanese	27	HIV
male	Malaysian	28	flu
female	American	35	HIV
female	Canadian	42	HIV
male	American	38	flu
male	Chinese	38	flu

- *Value Swapping:* Under value swapping [Reiss, 1984; Reiss et al., 1982], we can swap the two values (of the same attribute) of any two tuples in the dataset. Table 2.11 shows the anonymized table generated by value swapping. In this table, we swap Japanese and Malaysian for the first two tuples and swap 36 and 40 for the third and the fourth tuples.

Table 2.11: A published table by value swapping

Gender	Nationality	Age	Disease
male	Malaysian	26	HIV
male	Japanese	30	flu
female	American	40	HIV
female	Canadian	36	HIV
male	American	40	flu
male	Chinese	36	flu

The advantage of value swapping is that the domain of each *single* attribute after value swapping remains unchanged. In other words, all values in the table generated by value swapping must exist. However, after two values (of the same *single* attribute) of any two tuples are swapped, the combination of the swapped value of this attribute and the values of other attributes may not exist in one of these two tuples. For example, a combination of attributes (Gender, Occupation) = (male, waitress) is not possible since waitress corresponds to a female individual (although male exists in attribute Gender and waitress exists in attribute Occupation separately).

- *Model-Fitting-and-Regenerating:* Model-fitting-and-regenerating is a technique that involves the following three steps.

    * **Step 1 (Model Definition):** We define a data model with a number of parameters which can describe the model.
    * **Step 2 (Parameter Estimation):** We use data characteristics to estimate the parameters. Let $\mathcal{M}$ be the model described by these estimated parameters.

* **Step 3 (Data Regeneration):** Finally, we generate *another* set of data which follows model $\mathcal{M}$.

One representative method is *condensation* [Aggarwal and Yu, 2004] that defines a clustering model (in Step 1); estimates all clusters with their centers, radii and sizes (in Step 2); and generates another set of data according to these clusters (in Step 3).

Consider that each tuple in Table 2.2 can be represented as a single point. Without loss of generality, we illustrate them as 2-dimensional data points as shown in Figure 2.4(a). By using some clustering technique, we can find two clusters as shown in Figure 2.4(b). We can also obtain the cluster center, radius, and size of each cluster. After that, we re-generate all data points according to this information. The resulting data points are shown in Figure 2.4(c). Note that the six points in Figure 2.4(c) have different values than the six points in Figure 2.4(a), but they also form two clusters that are similar to the two original clusters in Figure 2.4(a). Finally, we can map back each point in Figure 2.4(c) to a corresponding tuple in a table.

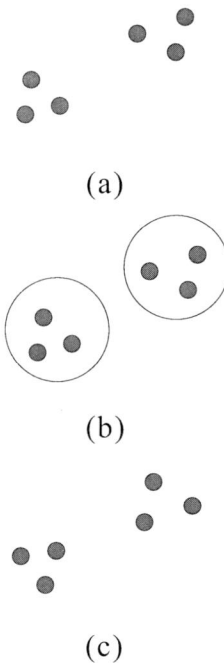

Figure 2.4: Condensation approach

Model-Fitting-and-Regenerating has the good feature that the statistics of the data captured by the model are maintained. For example, in the condensation approach, each cluster center, radius and size can be kept. However, similar to adding noise, it may generate some tuples that may not exist in real data.

2.2 INFORMATION LOSS METRIC

Since the data owner anonymizes the table, some information loss is inevitable. Consider the table generated by generalization as shown in Table 2.5. Some specific values are lost. Thus, we would like to quantify the *information loss* during anonymization. The information loss of a generalized table reduces the *utility* of the table; if the information loss is higher, then the utility is lower. There are a lot of different definitions on information loss in the literature. We describe one in this section and will discuss the others in Section 3.8.

We assume that each attribute is associated with a generalization taxonomy. An example can be found in Figure 2.3(a) that denotes the taxonomy of attribute Age. One way to define the cost of anonymization is based on this model. The cost is given by the *distortion ratio* of the resulting data set. If the value of the attribute of a tuple has not been generalized, there will be no distortion. However, if it is generalized, there is a distortion of the attribute of the tuple. If the value is further generalized (i.e., the original value is updated to a value at the node of the taxonomy near to the root), the distortion will increase. Thus, the *distortion* of this value is defined to be the *height* of the taxonomy from the original value to the updated value (or the height of the generalization of this value) divided by the total height of the taxonomy. For example, for a value that has not been generalized, the height of the generalization is equal to 0. Thus, the distortion is equal to 0. If the value is generalized one level up in the taxonomy, the height of the generalization is equal to $1/H$ where H is the height of the taxonomy. Let $d_{i,j}$ be the distortion of the value of attribute A_i of tuple t_j. The *distortion* of the whole data set is equal to the sum of the distortions of all values in the generalized data set. That is, distortion = $\sum_{i,j} d_{i,j}$. *Distortion ratio* is equal to the distortion of the generalized data set divided by the distortion of the *fully* generalized data set, where the fully generalized data set is one where all values of the attributes are generalized to the root of the taxonomy.

For example, consider the first tuple of Table 2.5. There is no distortion of male since there is no generalization of this value. However, the distortion of Asian is 1/2 because Chinese is generalized to Asian using the taxonomy (of height 2) shown in Figure 2.2(a). Similarly, according to the taxonomy as shown in Figure 2.3(a), the distortion of "26-30" is 1/2. Thus, the distortion for the first tuple is 0+1/2+1/2=1. By similar arguments, the distortion for each of the second, third, and fourth tuples is equal to 1, and the distortion for each of the last two tuples is equal to 1.5. The total distortion of the whole table is equal to 1 + 1 + 1 + 1 + 1.5 + 1.5 = 7.0. It is easy to verify that the distortion of the fully generalized table is equal to 18. Thus, the distortion ratio of this table is equal to 7.0/18 = 38.89%.

There are many other metrics to define the information loss due to the anonymization. Some metrics measure the information loss based on the anonymized table while others measure the accuracy of the results returned by some pre-defined query workloads over the anonymized table. Alternative metrics will be discussed in Section 3.8.

2.3 PRIVACY MODELS

We have discussed how a data table can be anonymized and how we measure the information loss of an anonymized table. In this section, we present some privacy models to quantify how to protect individual privacy.

Motivated by the privacy breach of an individual, *k-anonymity* has been proposed to protect individual privacy [Sweeney, 2002b].

Definition 2.1 *k*-**Anonymity.** A QI-group is said to satisfy *k-anonymity* (or a QI-group is said to be *k-anonymous*) if the size of the QI-group is at least k. A table T is said to satisfy *k-anonymity* (or a table is said to be *k-anonymous*) if each QI-group satisfies *k*-anonymity.

The objective of *k*-anonymity is to make sure that each individual is *indistinguishable* from at least $k-1$ other individuals in the table. Table 2.4 and Table 2.5 satisfy 2-anonymity. For example, consider Table 2.5. Although the adversary knows the QI attributes of Peter from the voter registration list, from this table, s/he cannot distinguish the first tuple from the second tuple. Thus, s/he is not sure whether Peter suffers from HIV because the first tuple is linked to HIV, but the second tuple is not.

As observed by LeFevre et al. [2005], *"for any anonymization mechanism, it is desirable to define some notion of minimality. Intuitively, a k-anonymization should not generalize, suppress, or distort the data more than it is necessary to achieve k-anonymity."* Thus, it is desirable to anonymize a table for *k*-anonymity with the minimum information loss. However, it has been shown that anonymizing a table for *k*-anonymity with the minimum information loss is NP-hard [Aggarwal et al., 2005].

Considering QI Attributes and Sensitive Attribute Together

As we discussed, the adversary cannot infer whether Peter suffers from HIV. However, even if the table satisfies 2-anonymity, the privacy protection of some individuals is insufficient.

Consider Table 2.5 again. Although the adversary cannot distinguish the third and the fourth tuples (which correspond to Mary and Sally, respectively), s/he is sure that both Mary and Sally suffer from HIV because both the third and the fourth tuples are linked to HIV in the table.

The major reason why *k*-anonymity cannot protect the individual privacy of Mary and Sally is that *k*-anonymity prevents an adversary from uniquely re-identifying an individual *without* considering the sensitive attribute Disease. Thus, other privacy models that consider both the QI attributes and the sensitive attributes are proposed. Some examples are *l*-diversity [Machanavajjhala et al., 2006], (α, k)-anonymity [Wong et al., 2006], and *t*-closeness [Li and Li, 2007].

In the literature, there are different interpretations of l-diversity. A simplified interpretation is given below.

Definition 2.2 l-**Diversity.** An QI-group is said to satisfy l-diversity (or an QI-group is said to be l-diverse) if the probability that any tuple in this group is linked to a sensitive value like HIV is at most $1/l$. The table satisfies l-diversity (or the table is said to be l-diverse) if each QI-group satisfies l-diversity.

For example, Table 2.12 is a 2-diverse table. There are two QI-groups. The first QI-group contains the first two tuples while the second QI-group contains the last four tuples. For each QI-group, the probability that a tuple is linked to HIV is at most 0.5.

Table 2.12: A 2-diverse published table by generalization

Gender	Nationality	Age	Disease
male	Asian	26-30	HIV
male	Asian	26-30	flu
*	Person	36-40	HIV
*	Person	36-40	HIV
*	Person	36-40	flu
*	Person	36-40	flu

Some more complicated forms of l-diversity are *entropy l-diversity* and *recursive (c, l)-diversity* [Machanavajjhala et al., 2006]. Their definitions can be found in Appendix A. However, due to their complicated forms, researchers commonly adopt the simplified interpretation of l-diversity given in Definition 2.2.

(α, k)-anonymity is one of the privacy models that consider both the QI attributes and the sensitive attribute. Given a real number $\alpha \in [0, 1]$ and a positive integer k, a table is said to satisfy (α, k)-anonymity (or a table is (α, k)-anonymous) if each QI-group in the table satisfies (α, k)-anonymity [Wong et al., 2006]. A QI-group G is said to satisfy (α, k)-anonymity if the number of tuples in G is at least k and the frequency (in fraction) of each sensitive value in G is at most α. Table 2.12 is an example of a $(0.5, 2)$-anonymous table. (α, k)-anonymity is similar to the simplified interpretation of l-diversity where α is set to $1/l$. However, (α, k)-anonymity considers additionally the k-anonymity requirement but l-diversity does not. Wong et al. [2006] show that anonymizing a table to satisfy (α, k)-anonymity with the mimimum information loss is NP-hard. When k is set to 1 and α is set to $1/l$, (α, k)-anonymity becomes the simplified model of l-diversity.

Since most privacy models (including the ones we discussed and some others that we will discuss later) share a common property, namely *monotonicity* property, the algorithms for generating a table under their models fall under a common framework that will be discussed in the next chapter.

Definition 2.3 Monotonicity. Let $\mathcal{R}$ be a privacy model. $\mathcal{R}$ is said to satisfy the *monotonicity* property if, for any two QI-groups G_1 and G_2 satisfying $\mathcal{R}$, the final QI-group that is a result of merging all tuples in G_1 and all tuples in G_2 satisfies $\mathcal{R}$.

For example, k-anonymity satisfies the monotonicity property. Take Table 2.5 for illustration. There are three QI-groups each of which satisfies 2-anonymity. Consider the first two QI-groups in the table (i.e., the QI-group containing the first two tuples and the QI-group containing the next two tuples). If we merge these two QI-groups into one single QI-group containing four tuples, the resulting QI-group also satisfies 2-anonymity. Furthermore, it is easy to verify that, for any two QI-groups in the original table, the QI-group that is a result of merging these two QI-groups also satisfies 2-anonymity.

Similarly, both l-diversity and (α, k)-anonymity satisfy the monotonicity property. Readers can verify this claim by examining Table 2.12.

Consider a microdata T and a privacy requirement $\mathcal{R}$ satisfying the monotonicity property. If a fully generalized table of T does not satisfy privacy requirement $\mathcal{R}$, then there does not exist any table T^* that is generalized from T such that T^* satisfies $\mathcal{R}$. For example, the fully generalized table (Table 2.6) of Table 2.2 does not satisfy 10-anonymity because there are only 6 tuples in the table and there are no sufficient tuples for 10-anonymity.

2.4 OTHER PRIVACY MODELS

In this section, we discuss alternative privacy models. As we described before, most privacy models follow the monotonicity property. In the following, for each privacy model, we first define the model and then check whether it satisfies the monotonicity property.

Numeric Sensitive Attribute

Our previous discussion assumed that the sensitive attribute is categorical. In some cases, the sensitive attribute can be numeric (e.g., Income). While a categorical sensitive attribute contains a limited number of possible values, a numeric sensitive attribute contains a large (or an infinite) number of possible values.

A straightforward approach to anonymizing data with a numeric sensitive attribute is to perform a discretization step over its values such that the numeric attribute is transformed to a categorical one, which can then be anonymized by using the techniques discussed in the previous section. However, this straightforward solution ignores the characteristics of numeric attributes, and thus this may increase information loss. In order to overcome these shortcomings, some alternative privacy models designed for data with numeric sensitive attributes are proposed.

Table 2.13: (k, e)-anonymity for $k = 2$ and $e = 5k$

Gender	Zipcode	Income
Male	54321	30k
Male	54321	30k
Female	54320	20k
Female	54320	10k
Female	54320	40k

(a) A microdata

Gender	Zipcode	Income
*	[54320-54321]	30k
*	[54320-54321]	30k
*	[54320-54321]	20k
*	[54320-54321]	10k
*	[54320-54321]	40k

(b) A (2, 5k)-anonymous table by global recoding

Gender	Zipcode	Income
*	[54320-54321]	30k
*	[54320-54321]	30k
*	[54320-54321]	20k
Female	54320	10k
Female	54320	40k

(c) A (2, 5k)-anonymous table by local recoding

- The first model is (k, e)-anonymity [Zhang et al., 2007], which generates a table where each QI-group is of size at least k and has a range of the sensitive values at least e. Consider the microdata table as shown in Table 2.13(a) (where Income is a sensitive numeric attribute). Table 2.13(b) and Table 2.13(c) are the anonymized tables satisfying $(2, 5k)$-anonymity by global recoding and local recoding, respectively. This is because, in each of these two tables, each QI-group is of size at least 2 and has a range on Income of at least $5k$.

However, (k, e)-anonymity does not consider the *proximity* of sensitive values in any QI-group, which may breach individual privacy. In a QI-group, some numeric sensitive values such as 20k are close to other numeric sensitive values such as 20.5k. If there are many close sensitive values (or even same sensitive values) in a QI-group, the adversary has high confidence to say that an individual in a QI-group is linked to these "close" sensitive values. For example, in Table 2.13(c), the first QI-group contains two occurrences of sensitive value 30k and only one occurrence of sensitive value 20k. The adversary can infer that each individual in this QI-group is linked to 30k with probability 2/3. Consider a more serious scenario where a

QI-group contains 999 sensitive values 30k and only one sensitive value 20k. Although this QI-group satisfies (2, 5k)-anonymity, the adversary has 99.9% confidence that an individual in this QI-group has salary 30k, which may breach individual privacy.

This privacy model (k, e)-anonymity satisfies the monotonicity property. Let us illustrate the monotonicity property with Table 2.13(c). which has two QI-groups each of which satisfies (2, 5k)-anonymity. If we merge these two QI-groups to form a new QI-group (Table 2.13(b)), it is easy to see that this new QI-group also satisfies (2, 5k)-anonymity.

Since (k, e)-anonymity satisfies the monotonicity property, we can adopt one of the existing algorithms described in Section 2.3.

- The second privacy model for data with the numeric sensitive attribute is called (ϵ, m)-anonymity [Li et al., 2008], which addresses the weakness of the (k, e)-anonymity that we just discussed. Here, ϵ is a non-negative real number and m is a positive integer. Intuitively, if the published table satisfies (ϵ, m)-anonymity, then each QI-group G must satisfy the following condition: For each sensitive numeric value that appears in G, the frequency (in fraction) of the tuples with sensitive numeric values close to s is at most $1/m$ where the closeness among numeric sensitive values is captured by parameter ϵ.

Li et al. [2008] propose two interpretations of (ϵ, m)-anonymity that capture the closeness among sensitive values:

  - The first interpretation is that the closeness of two numeric values, say s_1 and s_2, are captured by the absolute difference, $|s_1 - s_2|$, and, if the absolute difference between two numeric values is at most ϵ, then these two numeric values are defined to be close.
  - The second interpretation is related to the relative difference instead of the absolute difference. Specifically, a numeric value s_1 is close to s_2 if $|s_1 - s_2| \leq \epsilon s_2$.

Let $\epsilon = 5k$ and $m = 2$. Under the first interpretation, Table 2.13(a), that originally does not satisfy (ϵ, m)-anonymity, is anonymized to Table 2.14 satisfying (ϵ, m)-anonymity by global recoding. It is easy to verify that, in each QI-group of the anonymized table, for each sensitive value s_1, there are at most 1/2 tuples with sensitive value s_2 that have the absolute difference (i.e., $|s_1 - s_2|$) at most 5k. In this example, if we anonymize the table by local recoding, we also obtain the same anonymized table (Table 2.14).

However, (ϵ, m)-anonymity does not obey the monotonicity property. Let us illustrate with an example (Table 2.15(a)) by using the first interpretation. There are two QI-groups in Table 2.15(a) where each QI-group satisfies (10k, 2)-anonymity. Thus, the table satisfies (10k, 2)-anonymity. However, if we merge these QI-groups into a larger QI-group as shown in Table 2.15(b), then the large QI-group does not satisfy (10k, 2)-anonymity. In particular, there are three values close to value 20k whose absolute difference with 20k is at most 10k (10k, 20k and 30k). Note that there are totally four tuples in the QI-group. Since the frequency

Table 2.14: (ϵ, m)-anonymity for $\epsilon = 5k$ and $m = 2$

Gender	Zipcode	Income
*	[54320-54321]	30k
*	[54320-54321]	30k
*	[54320-54321]	20k
*	[54320-54321]	10k
*	[54320-54321]	40k

of the tuples with these values (i.e., 3/4) is more than 1/2, this QI-group violates (10k, 2)-anonymity. Thus, the merged QI-group violates (ϵ, m)-anonymity, which shows that it does not satisfy the monotonicity property.

Table 2.15: An example showing that (ϵ, m)-anonymity does not satisfy the monotonicity property

Gender	Zipcode	Income
Male	54321	10k
Male	54321	30k
Female	54320	20k
Female	54320	40k

(a) A table which satisfies (10k, 2)-anonymity

Gender	Zipcode	Income
*	[54320-54321]	10k
*	[54320-54321]	30k
*	[54320-54321]	20k
*	[54320-54321]	40k

(b) A table which does not satisfy (10k, 2)-anonymity

Personalized Privacy

Xiao and Tao [2006a] propose a personalized privacy model where each individual can provide his/her preference on the protection of his/her sensitive value, denoted by a *guarding node*. Consider the generalization taxonomy as shown in Figure 3.2. Table 2.16(a) shows the microdata where the QI attributes are Gender and Zipcode and the sensitive attribute is Education. An individual o with value "1st-4th" may specify "elementary" as a guarding node in order that any QI-group in the published table that may contain o should contain at most $1/l$ tuples with "elementary" values. This can be considered a variation of l-diversity. For $l = 2$, Table 2.16(b) and Table 2.16(c) are global and local recodings for Table 2.16(a), respectively.

Table 2.16: Anonymization for personalized anonymity

Gender	Zipcode	Education	Guarding Node
Male	54321	1st-4th	elementary
Female	54320	undergrad	none
Female	54320	undergrad	none

(a) A microdata

Gender	Zipcode	Education
*	[54320-54321]	1st-4th
*	[54320-54321]	undergrad
*	[54320-54321]	undergrad

(b) An anonymized table by global recoding

Gender	Zipcode	Education
*	[54320-54321]	1st-4th
*	[54320-54321]	undergrad
Female	54320	undergrad

(c) An anonymized table by local recoding

Since the personalized privacy model is based on l-diversity as the privacy requirement, it satisfies the monotonicity property.

Multiple Quasi-Identifier Attributes

In our discussions so far, we assumed that each individual can be uniquely identified using some external tables according to a *single* quasi-identifier. However, in some applications, individuals can be uniquely identified using *multiple* external tables according to *multiple* quasi-identifiers [Pei et al., 2009].

Let us adopt the example used by Pei et al. [2009] for illustration. Consider a microdata about road accidents in a region as shown in Table 2.17(a). Different people or parties can have different external tables about individuals. For example, as shown in Table 2.17(b), the auto insurance company has the vehicle registration records that contain the information of each driver about his/her name, age, vehicle and postcode. Note that the company does not have the occupation information because such information is not required in applying for auto insurance. At the same time, the human resource department has the resident records containing the information of each driver about his/her name, occupation, age and postcode, as shown in Table 2.17(c). Note that typically the human resource department has no information about residents' vehicles.

In this example, there are two quasi-identifiers, namely $QI_1 = \{$Age, Vehicle, Postcode$\}$ and $QI_2 = \{$Occupation, Age, Postcode$\}$. In general, there are multiple quasi-identifers, say $QI_1, QI_2, ..., QI_n$. Consider a privacy requirement $\mathcal{R}$ like k-anonymity. The objective of privacy preserving data publishing that considers multiple quasi-identifiers is to modify the original table

Table 2.17: Multiple quasi-identifiers

Occupation	Age	Vehicle	Postcode	Faulty
Dentist	30	Red Truck	31040	No
Family doctor	30	White Sedan	31043	Yes
Banker	30	Green Sedan	31043	No
Mortgage broker	30	Black Truck	31043	No

(a) A microdata about traffic accident records

Name	Age	Vehicle	Postcode
Alex	30	Red Truck	31040
Bob	30	White Sedan	31043
Chris	30	Green Sedan	31043
David	30	Black Truck	31043

(b) Vehicle registration records owned by the auto insurance company

Name	Occupation	Age	Postcode
Alex	Dentist	30	31040
Bob	Family doctor	30	31043
Chris	Banker	30	31043
David	Mortgage broker	30	31043

(c) Resident records owned by the human resource department

such that the modified table satisfies the privacy requirement $\mathcal{R}$ with respect to each quasi-identifier QI_i where $i \in [1, n]$.

Note that this privacy model with multiple quasi-identifier attributes can adopt one of the existing privacy requirements like k-anonymity for privacy protection. It is easy to verify that the model satisfies the monotonicity property.

Vague Boundary between Quasi-Identifier Attributes and Sensitive Attributes

We assume that the set of quasi-identifier attributes and the set of sensitive attributes in a microdata set are mutually exclusive. Wang et al. [2009] propose a privacy model called *free-form anonymity (FF anonymity)* in which some attributes of the microdata can be regarded as both quasi-identifer attributes and sensitive attributes. Specifically, some *values* of some attributes can be used as the quasi-identifier and some other *values* of these attributes can be regarded as sensitive.

The FF anonymity model is proposed based on whether a value is easily *observable*. If a value is easily observed, it is assumed that it is non-sensitive and is regarded as a quasi-identifier. Otherwise, it is regarded as a sensitive value.

Let us illustrate with the example shown in Table 2.18(a). Consider attribute Disease containing some sensitive values like HIV and some non-sensitive values like flu. Since flu and simple goiter can be *observed* on an individual, they are considered to be non-sensitive under *FF anonymity* privacy model. For example, a swelling of the neck of an individual indicates that s/he very likely suf-

Table 2.18: Illustration for the vague boundary between quasi-identifier attributes and sensitive attributes

Sex	Income	Disease
Female	250k	simple goiter
Male	250k	HIV
Male	150k	flu
Male	140k	HIV
Female	90k	flu

(a) A microdata

Name	Gender	Disease
Mary	Female	simple goiter

(b) External information about Mary

Name	Gender	Income
Alex	Male	High

(c) External information about Alex

fers from simple goiter. However, HIV cannot be observed easily and is thus considered as sensitive. If some values can be observed, these values can be used as some of the quasi-identifier attributes. For instance, suppose that the adversary knows that a female neighbor, Mary, suffers from simple goiter because of the swelling of her neck (as shown in Table 2.18(b)). With this information, if the adversary has Table 2.18(a), then the adversary can know that Mary has an income of 250k.

We discussed how Disease can be used as a quasi-identifier. In this example, Income can also be used for this purpose. An exact income is sensitive and cannot be observed by others easily. However, a range of incomes such as High, Medium and Low is non-sensitive and can be observed. For example, if Alex is a medical doctor, then the adversary may deduce that Alex's salary is high. Table 2.18(c) shows the information about Alex. According to Table 2.18(a) and Table 2.18(c), the adversary can infer that Alex suffers from HIV.

Any privacy requirement $\mathcal{R}$ can be applied on this privacy model after we define which values are regarded as quasi-identifiers and which values are regarded as sensitive values. Thus, one of the algorithms for $\mathcal{R}$ can still be adapted for this privacy model.

Publishing Additional Information for Better Utility

Kifer and Gehrke [2006] propose to publish some additional tables that are not sensitive at all so that these tables together can provide *better utility*. Consider the microdata shown in Table 2.1. Suppose that the privacy requirement is 2-diversity. In addition to publishing the 2-diverse table as shown in Table 2.12, in order to improve the utility of the published result(s), the data owner can publish three additional tables about the statistics of other attributes as shown in Table 2.19: a table about the statistics of attribute Gender (Table 2.19(a)), a table about the statistics of attribute Nationality (Table 2.19(b)) and a table about the statistics of attribute Age and attribute Disease

(Table 2.19(c)). Note that, in the 2-diverse table (Table 2.12), we do not have the detailed information about the attributes. In particular, in this table, we know that there are two male individuals and there are four individuals (which may be male or female). However, from the additional Table 2.19(a), we know that there are exactly four male individuals and two female individuals. Thus, more useful information about the statistics of attribute Gender can be obtained.

Table 2.19: Additional tables for better utility

Gender	Count
male	4
female	2

(a) A table about the statistics of attribute Gender

Nationality	Count
Japanese	1
Malaysian	1
American	2
Canadian	1
Chinese	1

(b) A table about the statistics of attribute Nationality

Age	Disease	Count
[21-30]	HIV	1
[21-30]	flu	1
[31-40]	HIV	2
[41-50]	flu	2

(c) A table about the statistics of attribute Age and attribute Disease

2.5 CONCLUSION

In this chapter, we presented the fundamental concepts that underlie all approaches to privacy preserving data publishing. In order to avoid a privacy breach, *the data publisher has to modify the data such that the modified data satisfy a privacy requirement, while, at the same time, minimizing the information loss of the modified data.*

We discussed various ways to modify the data, such as suppression, generalization, bucketization and perturbation. We also presented fundamental privacy requirements such as k-anonymity and l-diversity. Finally, we studied a metric to measure the information loss of the modified data.

The fundamental concepts described in this chapter will be used in techniques for one-time data publishing (Chapter 3) and multiple-time data publishing (Chapter 4). These concepts will also be adopted to other forms of data such as graph and spatial data (Chapter 5 and Chapter 6).

CHAPTER 3

One-Time Data Publishing

In Chapter 2, we discussed some fundamental concepts of data publishing. In this chapter, we describe the details of one-time data publishing based on these concepts.

As described in Section 1, an adversary could breach individual privacy when s/he has some background knowledge (Figure 1.1). In this chapter, we will see that different kinds of background knowledge may help the adversary breach individual privacy differently. Thus, in order to avoid a privacy breach, the data owner has to anonymize data accordingly.

3.1 KNOWLEDGE ABOUT QUASI-IDENTIFIERS

In Section 2.3, we described some well-known privacy models such as k-anonymity and l-diversity when an adversary has the knowledge about quasi-identifiers of individuals via an external table (e.g., a voter registration list). We also discussed that most privacy models share a common property, namely monotonicity property. Table 3.1 shows a summary stating whether a privacy model satisfies the monotonicity property. Some of the privacy models in the table will be described later in this chapter.

In the following, we describe some common techniques and approaches to anonymize a table if the privacy model satisfies the monotonicity property.

There are two major techniques used in the literature:

- Bottom-up Approach

- Top-down Approach

Bottom-up Approach

Consider the taxonomy in Figure 2.2(a). The least generalized value (or the most specific value) is located at the bottom while the most generalized value (or the least specific value) is located at the top.

The bottom-up approach [LeFevre et al., 2005; Samarati, 2001; Wang et al., 2004] generalizes the original table containing less general values to a table containing more general values. Specifically, it finds all QI-groups from the original table and then merges different QI-groups iteratively until each of the final merged QI-groups satisfies the desired privacy requirement. The final merged QI-groups correspond to the anonymized table to be published. Figure 3.1 shows the framework of the bottom-up approach that is specified in Algorithm 1.

Table 3.1: A summary stating whether a privacy model satisfies the monotonicity property

Privacy Model $\mathcal{R}$	Monotonicity?
k-anonymity [Sweeney, 2002b]	Yes
A simplified interpretation of l-diversity [Machanavajjhala et al., 2006]	Yes
Entropy l-diversity [Machanavajjhala et al., 2006]	Yes
Recursive l-diversity [Machanavajjhala et al., 2006]	Yes
(α, k)-anonymity [Wong et al., 2006]	Yes
(k, e)-anonymity [Zhang et al., 2007]	Yes
(ϵ, m)-anonymity [Li et al., 2008]	No
Personalized Privacy [Xiao and Tao, 2006a]	Yes
Privacy Model for Multiple Quasi-Identifier Attribute	Yes
FF-anonymity [Pei et al., 2009]	Yes
t-closeness [Li and Li, 2007]	Yes
(B, t)-privacy Li et al. [2009]	Yes
Privacy Model by Considering Knowledge about the Linkage of Individuals to Sensitive Values [Chen et al., 2007; Machanavajjhala et al., 2006; Tao et al., 2008]	Yes
m-confidentiality [Wong et al., 2009a]	Unknown
Injector [Li and Li, 2008]	Yes
Privacy Model by Considering Knowledge Mined from the Published Data [Kifer, 2009; Wong et al., 2009b]	Unknown

One typical bottom-up approach is the one adopted by Incognito algorithm [LeFevre et al., 2005] and Samarati's algorithm [Samarati, 2001]. Incognito was first proposed for k-anonymity, but it has also been applied in many other privacy requirements such as l-diversity [Machanavajjhala et al., 2006], t-closeness [Li and Li, 2007] and (α, k)-anonymity [Wong et al., 2006]. Let us focus on l-diversity as the privacy requirement for illustration.

Table 3.2 shows a data set containing two QI attributes (Gender and Education) and one sensitive attribute Disease where HIV is the sensitive value.

Figure 3.2 shows a taxonomy for Education. Figures 3.3(a) and (b) show the generalization domains of Gender and Education, respectively. Each node in a generalization taxonomy of attribute A corresponds to a *generalization domain* with respect to A. The generalization domain at the lower level has more detailed information than the domain at the higher level. For example, in Figure 3.3(a), generalization domain G0 (with respect to Gender) has more detailed information than G1. Domains of multiple attributes can be combined to more complex generalization domains such as $\langle G0, E1 \rangle$.

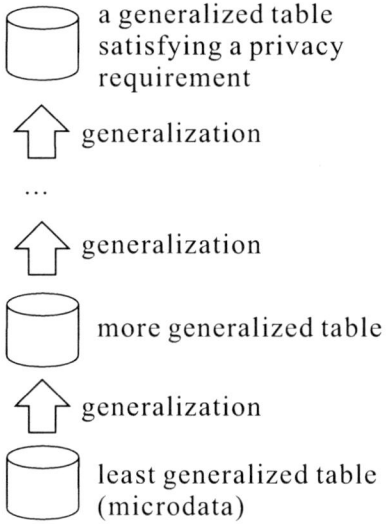

Figure 3.1: Bottom-up approach

Algorithm 1 Algorithm **Bottom-up**

Require: table T, privacy requirement $\mathcal{R}$
Ensure: table T^* that satisfies $\mathcal{R}$
1: fully generalize T to T'
2: **if** T' satisfies $\mathcal{R}$ **then**
3:   $T^* \leftarrow T$
4:   **while** T^* violates $\mathcal{R}$ **do**
5:     perform a generalization on T^*
6:   **end while**
7: **else**
8:   $T^* \leftarrow \emptyset$
9: **end if**
10: **return** T^*

Definition 3.1 Generalization Property. Let T be a table and let Q be a QI attribute set in T. Let G and G' be generalization domains with respect to Q, where G' is more general than G. Let T_G be the table that is generalized with the generalization domain G with respect to Q from T, and $T_{G'}$ be the table that is generalized with the generalization domain G' with respect to Q from T. If T_G is l-diverse, then $T_{G'}$ is also l-diverse.

For example, consider that the data owner wants to anonymize the data set in Table 3.2 for l-diversity. Let us set $l = 2$. Table 3.4, the table generalized with $\langle G1, E0 \rangle$, satisfies l-diversity. As

32 3. ONE-TIME DATA PUBLISHING

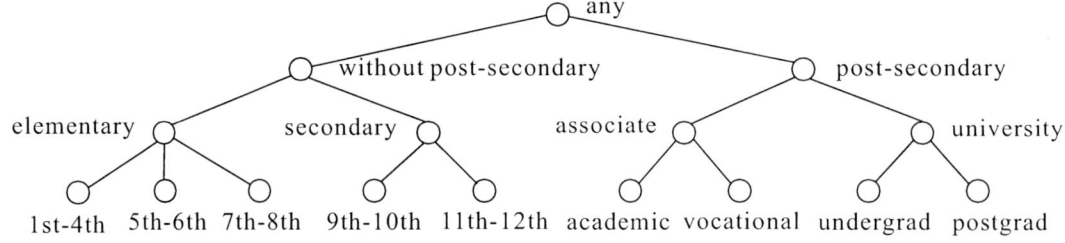

Figure 3.2: A taxonomy for attribute Education

Table 3.2: A data set

Gender	Education	Disease
Male	postgraduate	HIV
Female	undergraduate	HIV
Female	postgraduate	non-sensitive
Female	undergraduate	non-sensitive
Female	undergraduate	non-sensitive

Table 3.3: Table 3.2 anonymized with $\langle G0, E1 \rangle$

Gender	Education	Disease
Male	university	HIV
Female	university	HIV
Female	university	non-sensitive
Female	university	non-sensitive
Female	university	non-sensitive

$\langle G1, E1 \rangle$ is more general than $\langle G1, E0 \rangle$, the table generalized with domain $\langle G1, E1 \rangle$ should also be l-diverse (Table 3.5).

Let us describe an algorithm under this framework: First, all possible generalization domains (of multiple QI attributes) are generated. For example, Figure 3.4 shows a subset of all possible generalization domains. Then, we test whether the table generalized with a generalization domain G satisfies l-diversity from bottom to top. By the generalization property, if G satisfies l-diversity, we do not need to test generalization domains above it. This is because all generalization domains above it must satisfy l-diversity. Finally, the algorithm chooses the table that has the minimum information loss among all tables satisfying l-diversity.

For example, suppose our privacy requirement is 2-diversity and the information loss metric is the distortion ratio (defined in Section 2.2). We know that $\langle G0, E0 \rangle$ and $\langle G0, E1 \rangle$ do not satisfy

3.1. KNOWLEDGE ABOUT QUASI-IDENTIFIERS

(a)

G1 = {Person}
↑
G0 = {male, female}

(b)

E3 = {any}
↑
E2 = {without post-secondary, post-secondary}
↑
E1 = {elementary, secondary,..., university}
↑
E0 = {1st-4th, 5th-6th, ..., undergrad, postgrad}

Figure 3.3: Generalization domains of attributes Gender and Education

Table 3.4: Table 3.2 anonymized with $\langle G1, E0 \rangle$

Gender	Education	Disease
Person	postgraduate	HIV
Person	undergraduate	HIV
Person	postgraduate	non-sensitive
Person	undergraduate	non-sensitive
Person	undergraduate	non-sensitive

Table 3.5: Table 3.2 anonymized with $\langle G1, E1 \rangle$

Gender	Education	Disease
Person	university	HIV
Person	university	HIV
Person	university	non-sensitive
Person	university	non-sensitive
Person	university	non-sensitive

2-diversity but $\langle G1, E0 \rangle$ and $\langle G1, E1 \rangle$ satisfy 2-diversity. Since $\langle G1, E0 \rangle$ generates a 2-diverse table with less information loss, Table 3.4 (generalized with $\langle G1, E0 \rangle$) will be published.

Algorithm 2 shows the Incognito algorithm. Note that we shorten the description of the original algorithm by removing the part related to efficiency. However, the principle of the algorithm described here is the same.

Top-down Approach

The top-down approach [Fung et al., 2005; LeFevre et al., 2006] involves a *specialization* operation that is the inverse of generalization. Under generalization, a less general value like Chinese

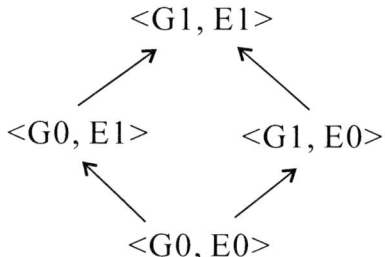

Figure 3.4: A subset of all possible generalization domains

Algorithm 2 Algorithm **Incognito**

Require: table T, privacy requirement $\mathcal{R}$
Ensure: table T^* that satisfies $\mathcal{R}$
1: generate all possible generalization domains of all QI attributes
2: **for** each generalization domain G **do**
3:     $T_G \leftarrow$ the table T that is generalized with G
4:     $IL_G \leftarrow$ the information loss of T_G (We can use any information loss metric. Please see Section 2.2 and Section 3.8 for details.)
5: **end for**
6: find the generalization domain G with the smallest information loss IL_G such that T_G satisfies $\mathcal{R}$
7: **if** such G exists **then**
8:     **return** T_G
9: **else**
10:     // there does not exist any feasible generalized table that satisfies $\mathcal{R}$
11:     **return** $\emptyset$
12: **end if**

---

is changed to a more general value like Asian. Under specialization, a more general value like Asian is modified to a less general value like Chinese.

The top-down approach first generalizes all attribute values in the table to the most general values. Then, one large QI-group is formed. The table is iteratively specialized while preserving the consistency of the original data values until the resulting specialized QI-groups violate the privacy requirement. The QI-groups formed in the second-to-last step correspond to the final anonymized table because all QI-groups in this step satisfy the privacy requirement. Figure 3.5 shows the framework of the top-down approach and Algorithm 3 specifies the top-down algorithm.

In the top-down approach, at each iteration, the data owner has to specialize a generalized table according to one attribute. This is because different attributes may yield different specializations as shown in Figure 3.6. For example, Table 3.5 can be specialized into two different tables according

3.1. KNOWLEDGE ABOUT QUASI-IDENTIFIERS 35

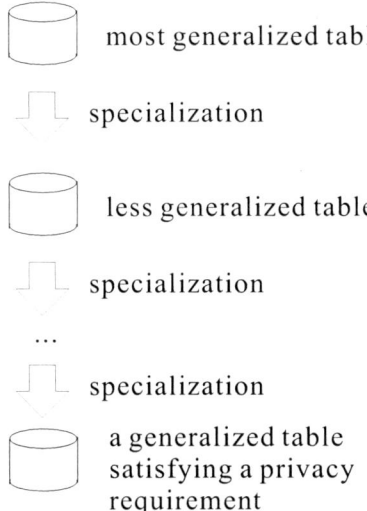

Figure 3.5: Top-down approach

Algorithm 3 Algorithm **Top-down**

Require: table T, privacy requirement $\mathcal{R}$
Ensure: table T^* that satisfies $\mathcal{R}$
1: fully generalize T to T^*
2: **if** T^* satisfies $\mathcal{R}$ **then**
3:    **while** T^* violates $\mathcal{R}$ **do**
4:       perform a specialization on T^*
5:    **end while**
6: **else**
7:    $T^* \leftarrow \emptyset$
8: **end if**
9: **return** T^*

---

to two attributes, Gender and Education. Table 3.3 and Table 3.4 are the tables obtained by specialization of Table 3.5 according to attribute Gender and attribute Education, respectively. In order to satisfy a privacy requirement, the data owner chooses one of the specialized tables that satisfies the privacy requirement. If there are multiple specialized tables that satisfy the privacy requirement, the top-down approach will choose one according to a heuristics function (e.g., the information loss of the specialized table).

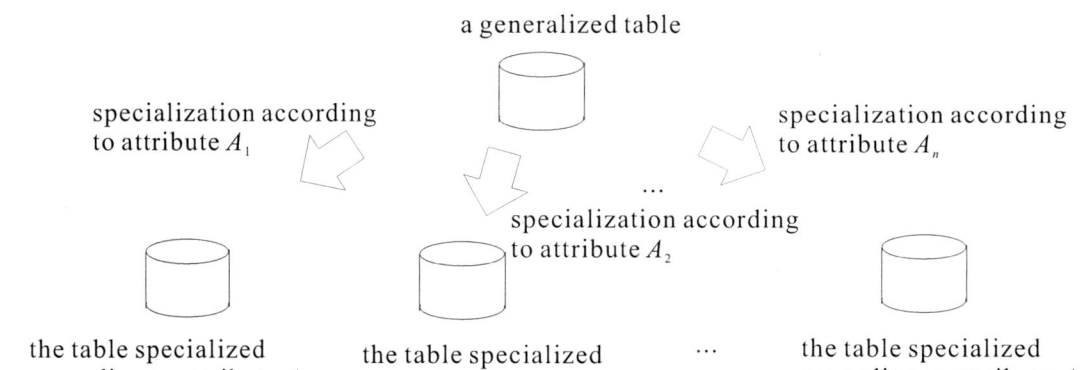

Figure 3.6: One step of specialization

Mondrian [LeFevre et al., 2006] is a top-down approach that we will discuss to illustrate the technique. Mondrian (see Algorithm 4) is a kd-tree based algorithm originally designed for k-anonymity; here, we discuss how it can be used for l-diversity.

Table 3.6: A raw table for the illustration of algorithm Mondrian			
Zipcode	Age	Disease
51102	36	HIV
51104	36	HIV
51101	31	non-sensitive
51102	31	non-sensitive
51104	31	non-sensitive
51105	31	non-sensitive

We illustrate the algorithm using Table 3.6 where the QI attributes are two numeric attributes, Zipcode and Age, and the sensitive attribute is Disease. Suppose the privacy requirement is 2-diversity. First, it fully generalizes the table as shown in Table 3.7 such that each tuple is in the same QI-group. Then, it iteratively chooses one attribute for partitioning the data; the iterations continue until there is no feasible partitioning.

Consider attribute Age for partitioning. The median of Age values is 31 (based on Table 3.6). Thus, range [31-36] is partitioned into range [31-31] (or value 31) and range [32-36]. The resulting table is shown in Table 3.8. However, it is easy to verify that Table 3.8 violates 2-diversity. So, the partitioning according to attribute Age is infeasible. Then, Mondrian chooses the remaining attribute (i.e., attribute Zipcode) for partitioning.

Table 3.7: A fully generalized table of Table 3.6

Zipcode	Age	Disease
[51101-51105]	[31-36]	HIV
[51101-51105]	[31-36]	HIV
[51101-51105]	[31-36]	non-sensitive
[51101-51105]	[31-36]	non-sensitive
[51101-51105]	[31-36]	non-sensitive
[51101-51105]	[31-36]	non-sensitive

Table 3.8: A table after the partition of Table 3.7 according to attribute Age

Zipcode	Age	Disease
[51101-51105]	[32-36]	HIV
[51101-51105]	[32-36]	HIV
[51101-51105]	31	non-sensitive
[51101-51105]	31	non-sensitive
[51101-51105]	31	non-sensitive
[51101-51105]	31	non-sensitive

Table 3.9: A table after the partition of Table 3.7 according to attribute Zipcode

Zipcode	Age	Disease
[51101-51103]	[31-36]	HIV
[51104-51105]	[31-36]	HIV
[51101-51103]	[31-36]	non-sensitive
[51101-51103]	[31-36]	non-sensitive
[51104-51105]	[31-36]	non-sensitive
[51104-51105]	[31-36]	non-sensitive

Consider attribute Zipcode for partitioning. Since the median value of attribute Zipcode is 51103, we obtain the resulting table as shown in Table 3.9. It is also easy to verify that Table 3.9 satisfies 2-diversity. Thus, Table 3.7 is partitioned to Table 3.9.

Mondrian then performs similar steps over each partition in Table 3.9. However, since there are no feasible partitionings such that each QI-group satisfies 2-diversity, Table 3.9 is published as the final output.

Although the example discussed above uses numeric QI attributes, Mondrian can also be applied to a table with the categorical QI attributes. In the example illustrated, the algorithm par-

Algorithm 4 Algorithm **Mondrian**

Require: table T, privacy requirement $\mathcal{R}$
Ensure: a number of partitions each of which satisfies $\mathcal{R}$
1: fully generalize T to T^*
2: **if** T^* satisfies $\mathcal{R}$ **then**
3:     $O \leftarrow$ **partition**(T^*) (See Algorithm 5)
4: **else**
5:     $O \leftarrow \emptyset$
6: **end if**
7: **return** O

Algorithm 5 Algorithm **partition**(T^*)

Require: table T^*, privacy requirement $\mathcal{R}$
Ensure: a number of partitions each of which satisfies $\mathcal{R}$

1: // try to check which attribute partitions the tuples such that each resulting partition satisfies privacy requirement $\mathcal{R}$
2: **for** each QI attribute A **do**
3:    $T' \leftarrow T^*$
4:    find the median $\bar{a}$ of attribute A in table T'
5:    $X_A \leftarrow$ a set of tuples in T' that have A values at most $\bar{a}$
6:    $Y_A \leftarrow$ a set of tuples in T' that have A values greater than $\bar{a}$
7:    **if** both X_A and Y_A satisfy privacy requirement $\mathcal{R}$ **then**
8:       // the two partitions according to attribute A satisfies $\mathcal{R}$
9:       $Satisfy_A \leftarrow$ **true**
10:      $IL_A \leftarrow$ the information loss of a table that contains two partitions X_A and Y_A (We can use any information loss metric. Please see Section 2.2 and Section 3.8 for details.)
11:    **else**
12:       $Satisfy_A \leftarrow$ **false**
13:    **end if**
14: **end for**
15: $R \leftarrow$ a set of attributes that have $Satisfy_A =$**true**
16: **if** $R = \emptyset$ **then**
17:    $O \leftarrow \{T^*\}$
18: **else**
19:    $A \leftarrow$ the attribute in R with the minimum information loss IL_A
20:    $P \leftarrow$ **partition**(X_A)
21:    $Q \leftarrow$**partition**(Y_A)
22:    $O \leftarrow P \cup Q$
23: **end if**
24: **return** O

titions the data into two parts according to the median of the numeric QI attributes. In the case of categorical QI attributes, the algorithm specializes each value to a number of less general values according to the generalization taxonomy. The remainder of the process is similar.

In the following sections, we will describe how the adversary makes use of other kinds of background knowledge to breach individual privacy, and how the data publisher can protect individual privacy accordingly. Since the background knowledge about the quasi-identifiers of individuals is fundamental and is easily accessible, in the following, when we consider one of the background knowledge, we assume that the adversary has the background knowledge about the quasi-identifiers of individuals without explicitly stating it.

3.2 KNOWLEDGE ABOUT THE DISTRIBUTION OF SENSITIVE VALUES

In the previous section, we have discussed that the adversary can use the knowledge about the quasi-identifier of individuals to breach individual privacy and have provided some mechanisms to protect the published data. In this section, we introduce another type of knowledge that is related to the distribution of values of the sensitive attribute. This is called *distribution-based background knowledge* [Li and Li, 2007; Li et al., 2009].

There are at least two kinds of distribution-based background knowledge:

- *Dataset-based distribution:* The dataset-based distribution is the distribution of the values of the sensitive attribute in the *entire* dataset [Li and Li, 2007]. Suppose that there are 100,000 individuals in the entire dataset T with 6,000 individuals linking to heart disease. The probability that an individual in the dataset is linked to heart disease is 0.06.

- *QI-based distribution:* The QI-based distribution is the distribution of the values in the sensitive attribute restricted to individuals with the same values on some QI attributes [Li et al., 2009]. Some well-known examples of such knowledge are the facts that Japanese seldom suffer from heart disease [Machanavajjhala et al., 2006] or male individuals cannot be linked to ovarian cancer [Li and Li, 2008]. In particular, the distribution of the sensitive attribute values of Japanese may be encoded as {(Japanese:"Heart Disease", 0.003), (Japanese:Flu, 0.21), ...} where (Japanese:x, p) denotes that the probability that a Japanese is linked to a value x is p. Table 3.10 shows the QI-based distribution of Nationality values.

Note that the dataset-based distribution is a special case of the QI-based distribution. If the group of individuals specified in the QI-based distribution is set to all individuals in the dataset, the QI-based distribution becomes the dataset based distribution. Thus, the QI-based distribution can be considered as the more powerful background knowledge.

Table 3.10: A QI based distribution of attribute Nationality

p()	Heart Disease	Flu	...
American	0.1	0.21	...
Japanese	0.003	0.22	...
Chinese	0.05	0.21	...
...	...	...	...

Dataset-Based Distribution

A privacy model called t-closeness [Li and Li, 2007] considers dataset based distribution. An QI-group is said to satisfy t-closeness if the distribution $\mathbb{P}$ of the sensitive values in this group is roughly equal to the distribution $\mathbb{Q}$ of the sensitive values in the whole table T.

Formally, suppose the sensitive attribute contains m values, namely $s_1, s_2, ..., s_m$. The distribution $\mathbb{P}$ of the sensitive values in an QI-group is defined to equal to $(p_1, p_2, ..., p_m)$ where p_i is the frequency (in fraction) of value s_i in the QI-group. The distribution $\mathbb{Q}$ of the sensitive values in the whole table is defined to be equal to $(q_1, q_2, ..., q_m)$ where q_i is the frequency (in fraction) of value s_i in the whole table. The difference between the distribution of the sensitive values in the QI-group ($\mathbb{P}$) and the distribution of the sensitive values in the whole table T ($\mathbb{Q}$), denoted as $D[\mathbb{P}, \mathbb{Q}]$, is at most t.

$$D[\mathbb{P}, \mathbb{Q}] = 1/2 \sum_{i=1}^{m} |p_i - q_i| \qquad (3.1)$$

A table T is said to satisfy t-closeness if each QI-group satisfies t-closeness.

Consider Table 3.11(a) where the QI attributes are Gender and Zipcode and the sensitive attribute is Disease. Table 3.11(a) does not satisfy 0.5-closeness. The reason is described as follows. There are two possible values of the sensitive attribute Disease. In the entire table T, there are 3 HIV values and 5 non-sensitive values. Let q_1 be the frequency (in fraction) of HIV in the whole table and q_2 be the frequency (in fraction) of a non-sensitive value in the whole table. The distribution $\mathbb{Q}$ of the sensitive values in the whole table is equal to $(q_1 q_2)$ where $q_1 = 3/8 = 0.375$ and $q_2 = 5/8 = 0.625$. Consider the original microdata as shown in Table 3.11(a). There is a QI-group with (Gender, Zipcode) = (Male, 54321). Let p_1 be the frequency (in fraction) of HIV in this QI-group and p_2 be the frequency (in fraction) of a non-sensitive value in this QI-group. The distribution $\mathbb{P}$ of the sensitive values in the QI-group is $(p_1 p_2)$ where $p_1 = 2/2 = 1$ and $p_2 = 0/2 = 0$. We have

$$\begin{aligned} D[\mathbb{P}, \mathbb{Q}] &= 1/2(|p_1 - q_1| + |p_2 - q_2|) \\ &= 1/2(|1 - 0.375| + |0 - 0.625|) \\ &= 0.625 \\ &> 0.5 \end{aligned}$$

Table 3.11: 0.5-closeness anonymization

Gender	Zipcode	Disease
Male	54321	HIV
Male	54321	HIV
Female	54320	non-sensitive
Female	54320	non-sensitive
Female	54320	non-sensitive
Female	54322	non-sensitive
Female	54322	non-sensitive
Female	54322	HIV

(a) A microdata

Gender	Zipcode	Disease
*	[54320-54322]	HIV
*	[54320-54322]	HIV
*	[54320-54322]	non-sensitive
*	[54320-54322]	non-sensitive
*	[54320-54322]	non-sensitive
*	[54320-54322]	non-sensitive
*	[54320-54322]	non-sensitive
*	[54320-54322]	HIV

(b) 0.5-closeness anonymization by global recoding

Gender	Zipcode	Disease
*	[54320-54321]	HIV
*	[54320-54321]	HIV
*	[54320-54321]	non-sensitive
*	[54320-54321]	non-sensitive
*	[54320-54321]	non-sensitive
Female	54322	non-sensitive
Female	54322	non-sensitive
Female	54322	HIV

(c) 0.5-closeness anonymization by local recoding

Thus, there exists a QI-group in Table 3.11(a) such that the difference between its distribution $\mathbb{P}$ and the distribution $\mathbb{Q}$ of the sensitive values in the whole table is greater than 0.5.

Table 3.11(b) and Table 3.11(c) show the anonymized tables satisfying 0.5-closeness by global recoding and local recoding, respectively. It is trivial to see that Table 3.11(b) satisfies 0.5-closeness since this table contains one QI-group that has the same distribution as the entire table. Let us elaborate why Table 3.11(c) satisfies 0.5-closeness. There are two QI-groups in this table. Consider the first QI-group with (Gender, Zipcode) = (*, [54320-54321]). The distribution $\mathbb{P}_1$ of the sensitive values in this group is (2/5, 3/5) = (0.4, 0.6). Thus,

$$\begin{aligned} D[\mathbb{P}_1, \mathbb{Q}] &= 1/2(|0.4 - 0.375| + |0.6 - 0.625|) \\ &= 0.025 \\ &< 0.5 \end{aligned}$$

Let $\mathbb{P}_2$ be the distribution of the sensitive values in the second QI-group with (Gender, Zipcode) = (Female, 54322). It is easy to verify that $D[\mathbb{P}_1, \mathbb{Q}] = 0.0417$ which is smaller than 0.5. Thus, the difference between the distribution of the sensitive values in each QI-group in this table and the distribution of the sensitive values in the whole table is at most 0.5. Table 3.11(c) satisfies 0.5-closeness.

It is easy to verify that the privacy requirement of t-closeness satisfies the monotonicity property (Definition 2.3). It should be noted that either the top-down approach or the bottom-up approach can be chosen for anonymizing the table in order to satisfy t-closeness.

QI-based Distribution

An example of the QI-based distribution of attribute `Nationality` is shown in Table 3.10. This distribution is said to be *certain* since all probabilities are known exactly.

Alternatively, the QI-based distribution can be *uncertain* [Li et al., 2009], which means that the adversary has the background knowledge about the QI-based distribution with some uncertainty. The uncertainty of the background knowledge is captured by an input parameter B that is a non-negative real number. The smaller the B value, the clearer is the background knowledge.

Li et al. [2009] propose a privacy model called (B, t)-*privacy* based on the QI-based distribution where B and t are two non-negative real numbers. Intuitively, B is an input parameter capturing the uncertainty of the background knowledge, and t is an input parameter controlling the privacy requirement.

Let T^* be the published table. Consider that an adversary wants to infer the sensitive value of an individual whose quasi-identifier value is q. Let $P_{pri}(B, q)$ be a prior belief about the distribution of the sensitive attribute of the individual with quasi-identifier value q before the adversary sees the published table T^*. Let $P_{pos}(B, q, T^*)$ be a posterior belief about the distribution of the sensitive attribute of the individual with quasi-identifier value q after the adversary sees the published table T^*. Intuitively, (B, t)-privacy protects privacy with the following rationale. Before the adversary sees the published table T^*, s/he has the prior belief $P_{pri}(B, q)$ about the distribution of the sensitive attribute of the individual with quasi-identifier q. After s/he sees T^*, s/he may change his/her belief to $P_{pos}(B, q, T^*)$ according to what s/he learnt from T^*. (B, t)-privacy guarantees that the change of the belief should not be significant.

A published table T^* is said to satisfy (B, t)-privacy if and only if the worst-case disclosure risk for all tuples (with QI value equal to q) is at most t:

$$\max_{q} D[P_{pri}(B, q), P_{pos}(B, q, T^*)] \leq t$$

where $D[\mathbb{P}, \mathbb{Q}]$ is the metric that measures the difference between distribution $\mathbb{P}$ and distribution $\mathbb{Q}$. One example of this metric is Equation (3.1).

Due to the uncertainty of background knowledge, Li et al. [2009] model both $P_{pri}(B, q)$ and $P_{pos}(B, q, T^*)$ by some kernel estimation methods. Consider that prior belief about background knowledge $P_{pri}(B, q)$ is *certain*, which can be illustrated in Table 3.10. Consider that the published table T^* contains one QI-group that has one Japanese and one American. In addition, this QI-group contains two values in the sensitive attribute, namely heart disease and flu. According to T^* alone, the adversary only thinks that American in this QI-group is linked to heart disease with 50% probability. However, with the additional background knowledge $P_{pri}(B, q)$, the adversary would change his/her belief and would think that American in this QI-group is linked to heart disease with probability higher than 50% because the Japanese in this QI-group has very less chance of linking to heart disease. Thus, the adversary obtains the posterior belief $P_{pos}(B, q, T^*)$ about the distribution of the sensitive attribute after s/he sees the published table T^*. The exact derivation of $P_{pos}(B, q, T^*)$ is given in [Li et al., 2009].

Intuitively, in most cases, the privacy model (B, t)-privacy satisfies the monotonicity property. Li et al. [2009] adopt approach Mondrian, one of the top-down approaches for the anonymization.

3.3 KNOWLEDGE ABOUT THE LINKAGE OF INDIVIDUALS TO SENSITIVE VALUES

A privacy breach can result if background knowledge is available linking some individuals to some sensitive values. There are two kinds of information:

- The information that some individuals do not have some sensitive values
- The information that some individuals have some sensitive values

3.3.1 INFORMATION THAT SOME INDIVIDUALS DO NOT HAVE SOME SENSITIVE VALUES

Both Machanavajjhala et al. [2006] and Chen et al. [2007] consider the background knowledge that some individuals do not have some sensitive values. For example, an individual who had mumps during his/her childhood is unlikely to suffer from mumps again [Martin et al., 2007]. Another example is that a male cannot suffer from ovarian cancer [Li and Li, 2008]. With this background knowledge, the adversary can breach individual privacy. Consider the published table as shown in Table 3.12(a). Attribute Name is used for illustration purpose only and is not disclosed in the published table. Assume that, according to the voter registration list, we know that the individuals in the first QI-group with (Gender, Zipcode) = (Male, [54321-54324]) are Peter and David. Similarly, Mary and Susan can also be found in the second QI-group with (Gender, Zipcode) = (Female, [54321-54324]). According to this table alone, the adversary knows that Peter suffers from HIV with 50% probability. However, suppose that the adversary knows that Peter cannot suffer from mumps in this published table. In this case, the adversary can easily deduce that Peter suffers from HIV with 100% probability.

Alternatively, the data owner can publish the table as shown in Table 3.12(b) by merging all tuples in Table 3.12(a) such that the probability that each individual is linked to a sensitive value at most 1/2. We say that this table satisfies 2-diversity since the major principle of 2-diversity discussed in Chapter 2 is still applied here. Consider the QI-group in Table 3.12(b) (that contains all tuples in the table). Even though we know that Peter cannot suffer from mumps, the adversary can deduce that Peter is linked to one out of three sensitive values, namely HIV, Cancer and flu; thus, the probability that Peter is linked to a sensitive value is equal to 1/3 (smaller than 1/2).

Table 3.12: An illustration of privacy breaches with background knowledge about whether some individuals are linked to some sensitive values

Name	Gender	Zipcode	Disease
Peter	Male	[54321-54324]	HIV
David	Male	[54321-54324]	Mumps
Mary	Female	[54321-54324]	Cancer
Susan	Female	[54321-54324]	flu

(a) A table where privacy is breached

Name	Gender	Zipcode	Disease
Peter	*	[54321-54324]	HIV
David	*	[54321-54324]	Mumps
Mary	*	[54321-54324]	Cancer
Susan	*	[54321-54324]	flu

(b) A table where no privacy is breached

We still have the monotonicity property for a privacy requirement $\mathcal{R}$ like l-diversity when we consider the knowledge that some individuals do not have some sensitive values. Thus, one of the existing algorithms can be adopted. For example, suppose that the adversary has the knowledge that an individual is not linked to a sensitive value. In particular, s/he knows that Peter cannot suffer from mumps. Table 3.13(a) satisfies 2-diversity since each QI-group satisfies 2-diversity. After we merge the two QI-groups in this table and obtain the resulting table as shown in Table 3.13(b), the resulting table also satisfies 2-diversity.

3.3.2 INFORMATION THAT SOME INDIVIDUALS HAVE SOME SENSITIVE VALUES

In addition to the information that some individuals do not have some sensitive values, the information that some individuals have some sensitive values can breach individual privacy [Chen et al., 2007; Tao et al., 2008]. For example, an adversary can obtain the diagnostic results of some patients

Table 3.13: An example illustrating the monotonicity property by considering the knowledge about the linkage of individuals to sensitive values

Name	Gender	Zipcode	Disease
Bob	Male	[54321-54324]	flu
Peter	Male	[54321-54324]	HIV
David	Male	[54321-54324]	Mumps
Mary	Female	[54321-54324]	Cancer
Susan	Female	[54321-54324]	flu
Grace	Female	[54321-54324]	fever

(a) A table satisfying 2-diversity

Name	Gender	Zipcode	Disease
Bob	*	[54321-54324]	flu
Peter	*	[54321-54324]	HIV
David	*	[54321-54324]	Mumps
Mary	*	[54321-54324]	Cancer
Susan	*	[54321-54324]	flu
Grace	*	[54321-54324]	fever

(b) Another table satisfying 2-diversity

via a friend working in the hospital [Tao et al., 2008]. Another example is that, being an adversary, the boss of a company can read the sick-leave records of his employees that show their diseases.

Similarly, the adversary can breach the privacy of *other* individuals. The idea is similar. In the published table as shown in Table 3.12, if the adversary knows that David suffers from mumps, s/he can deduce that Peter suffers from HIV with 100% probability.

Similarly, we still have the monotonicity property when the adversary has. the information that some individuals have some sensitive values.

3.4 KNOWLEDGE ABOUT THE RELATIONSHIP AMONG INDIVIDUALS

The knowledge about the relationship among individuals can be used to breach individual privacy. One typical example is the knowledge that, if an individual is linked to HIV (a sensitive value), his wife may also be linked to HIV [Martin et al., 2007].

Martin et al. [2007] and Chen et al. [2007] consider the background knowledge about the relationship among individuals. The background knowledge defined by Martin et al. [2007] is a set of *k implications* each of which has the following form: if individual p_1 is linked to sensitive value v_1, then individual p_2 is linked to sensitive value v_2, where v_1 may equal v_2. This model assumes that an adversary can equip with the background knowledge containing a set of k implications Individual privacy is breached if the probability that an individual is linked to a sensitive value given this background knowledge is greater than a privacy threshold c where c is a user parameter. Since the data publisher does not know which k implications the adversary has, the model considers all possible background knowledge for privacy protection. Specifically, the model requires that the probability that *any* individual is linked to a sensitive value given the background knowledge containing *any* k implications is at most c. Similar implications are proposed by Chen et al. [2007], but v_1 is restricted to be equal to v_2.

Consider Table 3.12. As we illustrated, after the adversary has the background knowledge that David suffers from mumps, s/he can deduce that Peter suffers from HIV. If the adversary is equipped with one additional background knowledge that Peter and Mary are a couple, then the adversary can have one more inference that Mary may also suffer from HIV.

Intuitively, the privacy model by considering the knowledge about the relationship among individual satisfies the monotonicity property and thus one of the existing algorithms can be adopted.

3.5 KNOWLEDGE ABOUT ANONYMIZATION

In the previous discussion, we did not consider that the adversary can know the anonymization mechanism. Now, we study the case where the adversary has some additional knowledge about the mechanism involved in the anonymization, and it launches an attack based on this knowledge. We use l-diversity to illustrate the issue.

Table 3.14(a) shows a microdata table where QI is the quasi-identifier attributes and Disease is the sensitive attribute containing only one sensitive value, HIV. Suppose the QI values of $q1$ and $q2$ can be generalized to Q. For example, $q1$ may be {*Nov* 1930, Z3972, *M*}, $q2$ may be {*Dec* 1930, Z3972, *M*} and Q is {*Nov/Dec* 1930, Z3972, *M*}[1]. Recall that a tuple associated with a sensitive value such as HIV is called a *sensitive* tuple. For each QI-group, at most half of the tuples are sensitive. Hence, the table satisfies 2-diversity.

As described in Section 2.3, existing approaches of anonymization for data publishing have an implicit principle [LeFevre et al., 2005]: *"For any anonymization mechanism, it is desirable to define some notion of minimality. Intuitively, a k-anonymization should not generalize, suppress, or distort the data more than it is necessary to achieve k-anonymity."* Based on this minimality principle, Table 3.14(a) will not be generalized.[2] In fact, the above notion of minimality is too strong since almost all known anonymization problems for data publishing are NP-hard; many existing algorithms are heuristic and usually attain local minima. We shall later give a more relaxed notion of the minimality principle in order to cover both the optimal as well as the heuristic algorithms. For now, we assume that minimality principle implies that a QI-group will not be generalized unnecessarily.

Next, consider a slightly different table, Table 3.14(b). Here, the set of tuples for $q1$ violates 2-diversity because the proportion of the sensitive tuples is greater than $1/2$. Thus, this table will

---
[1] Note that $q1$ and $q2$ may themselves be generalized values.
[2] This is the case for each of the anonymization algorithms proposed by Machanavajjhala et al. [2006], Wang et al. [2007] and Wong et al. [2006].

Table 3.14: 2-diversity: global and local recoding

QI	Disease
$q1$	HIV
$q1$	non-sensitive
$q2$	HIV
$q2$	non-sensitive
$q2$	non-sensitive
$q2$	non-sensitive
$q2$	non-sensitive

(a) good table

QI	Disease
$q1$	HIV
$q1$	HIV
$q2$	non-sensitive
$q2$	non-sensitive
$q2$	non-sensitive
$q2$	non-sensitive
$q2$	non-sensitive

(b) bad table

QI	Disease
Q	HIV
Q	HIV
Q	non-sensitive
Q	non-sensitive
Q	non-sensitive
Q	non-sensitive
Q	non-sensitive

(c) global

QI	Disease
Q	HIV
Q	HIV
Q	non-sensitive
Q	non-sensitive
$q2$	non-sensitive
$q2$	non-sensitive
$q2$	non-sensitive

(d) local

be anonymized to a *generalized* table by generalizing the QI values as shown in Table 3.14(c) by *global recoding* [Wang and Fung, 2006; Xiao and Tao, 2006a]. The anonymization by local recoding is shown in Table 3.14(d). These anonymized tables satisfy 2-diversity. The question we are interested in is whether these tables really protect individual privacy.

Most commonly, the knowledge of the adversary involves an external table T^e such as a voter registration list that maps QIs to individuals [LeFevre et al., 2005, 2006; Sweeney, 2002b; Xiao and Tao, 2006a]. It is commonly assumed that each tuple in T^e maps to one individual and no two tuples map to the same individual. The same is also assumed in the table T to be published. Let us first consider the case when T and T^e are mapped to the same set of individuals. Table 3.15(a) is an example of T^e.

Assume further that the adversary knows the goal of 2-diversity, s/he also knows whether it is a global or local recoding, and Table 3.15(a) is available as the external table T^e. With the notion of minimality in anonymization, the adversary reasons as follows: In the published Table 3.14(c), there are 2 sensitive tuples. From T^e, there are 2 tuples with QI=$q1$ and 5 tuples with QI=$q2$. Hence, the QI-group for $q2$ in the original table *must* already satisfy 2-diversity because even if both sensitive tuples have QI=$q2$, the proportion of sensitive values in the group for $q2$ is only 2/5. Since *generalization* has taken place, at least one QI-group in the original table T must have violated 2-diversity, because otherwise no generalization will take place according to minimality. The adversary

Table 3.15: T^e: external table available to the adversary

Name	QI
Andre	$q1$
Kim	$q1$
Jeremy	$q2$
Victoria	$q2$
Ellen	$q2$
Sally	$q2$
Ben	$q2$

(a) individual QI

QI
$q1$
$q1$
$q2$
$q2$
$q2$
$q2$
$q2$

(b) multiset

Name	QI
Andre	$q1$
Kim	$q1$
Jeremy	$q2$
Victoria	$q2$
Ellen	$q2$
Sally	$q2$
Ben	$q2$
Tim	$q4$
Joseph	$q4$

(c) individual QI

QI
$q1$
$q1$
$q2$
$q2$
$q2$
$q2$
$q2$
$q4$
$q4$

(d) multiset

concludes that $q1$ has violated 2-diversity, and that is possible only if both tuples with QI=$q1$ have Disease attribute value of HIV. The adversary therefore discovers that Andre and Kim are linked to HIV.

In some works, it is assumed that the set of individuals in the external table T^e can be a superset of those for the published table. Table 3.15(c) shows such a case where there are no tuples for Tim and Joseph in Table 3.14(a) and Table 3.14(b). If it is known that $q4$ cannot be generalized to Q (e.g., $q4=\{Nov\ 1930, Z3972, F\}$ and $Q=\{Jan/Feb\ 1990, Z3972, M\}$), then the adversary can be certain that the tuples with QI=$q4$ are not in the original table. Thus, the tuples with QI=$q4$ in T^e do not have any effect on the above reasoning of the adversary, and, therefore, the same conclusion that Andre and Kim are linked to HIV can be drawn. Such an attack based on the minimality principle is called a *minimality attack*.

Observation 3.2 If a table T is anonymized to T^* that satisfies l-diversity, it can suffer from a minimality attack. This is true for both global and local recoding and for the cases when the set of individuals related to T^e is a superset of those related to T.

Table 3.16: 2-diversity (where all values in Disease are sensitive): global and local recoding

QI	Disease
$q1$	HIV
$q1$	Lung Cancer
$q2$	Gallstones
$q2$	HIV
$q2$	Ulcer
$q2$	Alzheimer
$q2$	Diabetes
$q4$	Ulcer
$q4$	Alzheimer

(a) good table

QI	Disease
$q1$	HIV
$q1$	HIV
$q2$	Gallstones
$q2$	Lung Cancer
$q2$	Ulcer
$q2$	Alzheimer
$q2$	Diabetes
$q4$	Ulcer
$q4$	Alzheimer

(b) bad table

QI	Disease
Q	HIV
Q	HIV
Q	Gallstones
Q	Lung Cancer
Q	Ulcer
Q	Alzheimer
Q	Diabetes
$q4$	Ulcer
$q4$	Alzheimer

(c) global

QI	Disease
Q	HIV
Q	HIV
Q	Gallstones
Q	Lung Cancer
$q2$	Ulcer
$q2$	Alzheimer
$q2$	Diabetes
$q4$	Ulcer
$q4$	Alzheimer

(d) local

In the above example, some values of the sensitive attribute `Disease` are not sensitive. For example, would it help the adversary breach individual privacy if all values in the sensitive attributes are sensitive? In the tables in Table 3.16, we assume that all values for `Disease` are sensitive. Table 3.16(a) satisfies 2-diversity but Table 3.16(b) does not. Suppose anonymization of Table 3.16(b) results in Table 3.16(c) by global recoding and Table 3.16(d) by local recoding. An adversary who is armed with the external table Table 3.15(c) and the knowledge that data are anonymized using 2-diversity can launch an attack by reasoning as follows: with 5 tuples for QI=$q2$ and each sensitive value appearing at most twice, there cannot be any violation of 2-diversity for the tuples with QI=$q2$. There must have been a violation for QI=$q1$. For a violation to take place, both tuples with QI=$q1$ must be linked to the same disease. Since HIV is the only disease that appears twice, Andre and Kim must have contracted HIV.

Observation 3.3 Minimality attack is possible whether or not the sensitive attribute contains non-sensitive values.

3.5. KNOWLEDGE ABOUT ANONYMIZATION

Recall that the intended *objective* of 2-diversity is to make sure that an adversary cannot deduce with a probability above 1/2 that an individual is linked to any sensitive value. Thus, the published tables violate this objective.

Wong et al. [2009a] propose a privacy requirement called *m-confidentiality* that guarantees to protect individual privacy taking into consideration that the adversary has the knowledge about the anonymization. It is not easy to see whether this privacy requirement satisfies the monotonicity property. Recall that all algorithms described in Section 2.3 rely on the monotonicity property and the minimality principle. Thus, m-confidentiality is much more complicated than other privacy requirements discussed before. This motivates Wong et al. [2009a] to propose another algorithm called called *MASK* (Minimality Attack Safe K-anonymity) that enforces m-confidentiality. This algorithm involves two major steps (Algorithm 6):

- **Step 1 (k-anonymization Step):** The first step is to anonymize the table such that the anonymized table satisfies k-anonymity. Any existing techniques for k-anonymity can be used in this step.

- **Step 2 (Precaution Step):** An additional precaution step is executed to distort some values in the sensitive attribute for further privacy protection.

Specifically, in Step 2 of algorithm MASK, Wong et al. [2009a] make use the concept of l-diversity to generate a table which satisfies l-diversity as an intermediate step. With this table, some values are distorted such that the resulting table satisfies m-confidentiality. Note that the original l-diversity privacy model does not consider the minimality principle but m-confidentiality does.

Consider Algorithm 6. After Step 1, some QI-groups may not satisfy l-diversity. Steps 2(a) to 2(c) above will ensure that all QI-groups in the result are l-diverse. In Step 2(a), we select a QI-group set $\mathcal{L}$ from T^k. The purpose is to disguise the distortion so that the adversary cannot tell the difference between a distorted QI-group and many un-distorted QI-groups.

We illustrate algorithm MASK with the raw table as shown in Table 3.17(a). Suppose we set $k = 2$ and $m = 2$. The first step is to generate a 2-anonymous table. Suppose $q1$ and $q2$ can be generalized to Q. Firstly, we adopt an algorithm for k-anonymity that generates table T^k (Table 3.17(b)) where all QI-groups are of size at least 2. Secondly, from Table 3.17(b), we obtain $\mathcal{V} = \{Q\}$. This is because the set of tuples with QI=Q violates 2-diversity. We also obtain $\mathcal{L} = \{q3\}$. This is because both the set of tuples with QI=$q3$ and the set of tuples with QI=$q4$ satisfy 2-diversity. More specifically, the proportion of the tuples with QI=$q3$ is 0.5 and the proportion of the tuples with QI=$q4$ is 0. Since we set the size of $\mathcal{L}$ to be equal to $(l-1) \times |\mathcal{V}| = (2-1) \times 1 = 1$, we choose the QI-group with QI=$q3$ for $\mathcal{L}$ because it has the greatest proportion of HIV among the QI-groups in T^k that satisfy 2-diversity. Thirdly, we determine the distribution $\mathcal{D}$ of the p_i values. Since there is only one QI-group (with HIV proportion equal to 0.5) in $\mathcal{L}$, we create a distribution $\mathcal{D}$ such that 0.5 must be returned whenever we draw from distribution $\mathcal{D}$. (Suppose $\mathcal{L} = \{q3, q4\}$. Since the proportions of the QI-group with QI=$q3$ and the QI-group with QI=$q4$ are equal to 0.5 and 0, respectively, $\mathcal{D}$ is the distribution where the probability that 0.5 is returned is equal to 0.5 and the probability that

Algorithm 6 Algorithm **MASK**

1: **Step 1:** From the given table T, generate a k-anonymous table T^k where k is a user parameter.

2: **Step 2(a):** From T^k, determine the set $\mathcal{V}$ containing all QI-groups that violate l-diversity in T^k, and a set $\mathcal{L}$ containing QI-groups that satisfy l-diversity in T^k.

   (a) We set the size of $\mathcal{L}$, denoted by u, to $(l-1) \times |\mathcal{V}|$. If the total number of QI-groups that satisfy l-diversity in T^k is smaller than $u(=(l-1) \times |\mathcal{V}|)$, we report that the table cannot be published. (Note: This case is rare because, typically, there are rare sensitive values in a table, and there are a sufficient number of QI-groups that satisfy l-diversity in T^k.) Otherwise, we do the following: among all the QI-groups in T^k that satisfies l-diversity, we pick u QI-groups with the highest proportions of the sensitive value set s and insert them into $\mathcal{L}$.

   (b) If $\mathcal{V} = \emptyset$, then we can return T^k as our published table. Otherwise, then we do continue the following steps.

3: **Step 2(b):** For each QI-group Q_i in $\mathcal{L}$, find the proportion p_i of tuples containing values in the sensitive value set s. The distribution $\mathcal{D}$ of the p_i values is determined.

4: **Step 2(c):** For each QI-group $E \in \mathcal{V}$, randomly pick a value of p_E from the distribution $\mathcal{D}$. The sensitive values in E are distorted in such a way that the resulting proportion of the sensitive value set s in E is equal to p_E. We name the QI-group E in which sensitive values are distorted as a *distorted QI-group*.

---

0 is returned is equal to 0.5.) Fourthly, we randomly pick a value from distribution $\mathcal{D}$. Suppose 0.5 is picked. Since $\mathcal{V}$ contains the QI-group E with QI=Q only, the sensitive values in E are distorted such that the resulting proportion of HIV in E is equal to 0.5. For example, we change one of the sensitive values in E to a non-sensitive value. Finally, Table 3.17(c) is generated and is ready to be published.

3.6 KNOWLEDGE MINED FROM THE MICRODATA

In the above discussion, in order to protect individual privacy in the published data, the data owner has to *explicitly* specify the background knowledge the adversary may have. It is possible to mine the background knowledge from the microdata automatically [Li and Li, 2008]. The rationale is that, if some fact exists in the real world, the fact must also exist in the microdata. One fact is that male individuals cannot be linked to ovarian cancer. This fact must exist in the microdata because we cannot find a male individual who suffers from ovarian cancer.

Specifically, Li and Li [2008] focus on mining negative association rules in the form of $X \rightarrow \neg Y$ as the background knowledge that can be mined from the microdata, where X is a predicate

Table 3.17: An illustration of algorithm MASK

QI	Disease
$q1$	HIV
$q2$	HIV
$q3$	HIV
$q3$	non-sensitive
$q4$	non-sensitive
$q4$	non-sensitive

(a) A raw table T

QI	Disease
Q	HIV
Q	HIV
$q3$	HIV
$q3$	non-sensitive
$q4$	non-sensitive
$q4$	non-sensitive

(b) 2-anonymous table of T

QI	Disease
Q	HIV
Q	non-sensitive
$q3$	HIV
$q3$	non-sensitive
$q4$	non-sensitive
$q4$	non-sensitive

(c) 2-confidential table of T

involving only some of the quasi-identifier attributes and Y is a predicate involving only the sensitive attribute. "Male $\rightarrow \neg$ Ovarian Cancer" is an example that can be mined from the microdata. Suppose the data owner modifies the microdata as shown in Table 3.18 to the 2-diverse table as shown in Table 3.19 by bucketization. Consider the QI-group with GID = G_1 that contains one male and one female. Note that there is only one occurrence of ovarian cancer in this group. Using this negative association rule, the adversary can deduce that the female (Mary) in this QI-group suffers from ovarian cancer.

Consider a privacy requirement $\mathcal{R}$ satisfying the monotonicity property. Negative association rules can be considered as the knowledge that some individuals do not have some sensitive values. Similar to Section 3.3, by considering the negative association rules, $\mathcal{R}$ also satisfies the monotonicity property. Thus, one of the algorithms described in Section 2.3 can be used.

Table 3.18: An example

Name	Sex	Zipcode	Disease
Alex	male	54321	Heart Disease
Mary	female	54323	Ovarian Cancer
Clement	male	54322	Flu
David	male	54327	Stomach Virus
Emily	female	32134	HIV
Grace	female	32135	Ovarian Cancer
Helen	female	32138	flu
Iris	female	32139	fever
...	...	...	...

Table 3.19: A 2-diverse dataset anonymized from Table 3.18

Sex	Zipcode	GID
male	54321	G_1
female	54323	G_1
male	54322	G_2
male	54327	G_2
female	32134	G_3
female	32135	G_3
female	32138	G_4
female	32139	G_4
...	...	...

(a) QI Table

GID	Disease
G_1	Heart Disease
G_1	Ovarian Cancer
G_2	Flu
G_2	Stomach Virus
G_3	HIV
G_3	Ovarian Cancer
G_4	flu
G_4	fever
...	...

(b) Sensitive table

However, directly adopting one of these algorithms may generate a table with high information loss. For example, algorithm Incognito is a global recoding algorithm and adopting it for this privacy requirement may generate a table with high information loss.

Algorithm Injector [Li and Li, 2008] addresses this in two major steps:

- **Step 1 (Negative Association Rule Mining):** All negative association rules are mined using some data mining technique.

- **Step 2 (Data Anonymization):** These association rules are used for data anonymization.

The negative association rules suggest that some quasi-identifier attributes cannot be put together into one single QI-group that contains some sensitive values. An efficient algorithm is proposed to generate the published data protecting individual privacy according to all the mined negative association rules.

Specifically, we can use any existing algorithm (e.g., [Agrawal and Srikant, 1994]) for mining all negative association rules for Step 1. Before we describe the details of Step 2, we first give some definitions that will be used in the algorithm.

Assume that the privacy requirement $\mathcal{R}$ is l-diversity. We say that a tuple *cannot take* a sensitive value s if there exists a negative association rule $X \rightarrow \neg s$ and the tuple has quasi-identifier attribute values equal to X. For example, since we know "Male $\rightarrow \neg$ Ovarian Cancer", in Table 3.18, a tuple for male (e.g., Alex) cannot take Ovarian Cancer.

Consider the QI-group with GID = G_2 in Table 3.19. From the perspective of an adversary, there are two possible *assignments*. The first assignment is that Clement's tuple is linked to flu and David's tuple is linked to stomach virus. The second assignment is that Clement's tuple is linked to stomach virus and David's tuple is linked to flu. Since Clement's tuple can be linked to two sensitive values, namely flu and stomach virus, we say that Clement's tuple has 2 *valid* sensitive values in this QI-group with GID = G_2.

Consider the QI-group with GID = G_1 in the same table. There is only one possible assignment because we know "Male $\rightarrow \neg$ Ovarian Cancer". The assignment is that Alex's tuple is linked to heart disease and Mary's tuple is linked to ovarian cancer. In this case, we say that Alex's tuple has 1 *valid* sensitive value in this QI-group, namely heart disease.

Consider a QI-group containing N tuples and N sensitive values. Formally, given a tuple t and a sensitive value s in this group, s is said to be *valid* for tuple t if there exists a possible assignment between tuples and sensitive values such that t is linked to s. Note that ovarian cancer is not a valid sensitive value for Alex's tuple in the QI-group with GID = G_1. This is because there is no assignment that Alex, a male individual, is linked to ovarian cancer provided that we know "Male $\rightarrow \neg$ Ovarian Cancer".

Consider two tuples t and t' where t has a sensitive value s and t' has a sensitive value s'. Tuple t is said to be *incompatible* with tuple t' if at least one of the following three conditions holds:

- $s = s'$
- t cannot take value s', and
- t' cannot take value s

For example, in Table 3.18, Grace's tuple is incompatible with Mary's tuple because both tuples are linked to the same sensitive value, ovarian cancer. Besides, Alex's tuple is incompatible with Mary's tuple because Alex's tuple cannot take ovarian cancer, Mary's sensitive value. However, Mary's tuple is compatible with Emily's tuple because they have unequal sensitive values, Mary's tuple can take value HIV, and Emily's tuple can take Ovarian Cancer.

A tuple t is said to be incompatible with a QI-group G if t is incompatible with at least one tuple in G. For instance, since Grace's tuple is incompatible with Mary's tuple, Grace's tuple is incompatible with the QI-group with GID = G_1 containing Mary's tuple.

Now, we are ready to describe the detailed sub-steps in Step 2. There are two phases for Step 2.

- **Phase 1 (Initial Grouping Phase):** Firstly, we find a tuple t that has the largest number of incompatible tuples. We want to find some other tuples that can be merged with t such that each tuple in the QI group formed by these tuples has l valid sensitive values. If we cannot find such a QI-group, we insert t into a new table called T_r that will be processed in Phase 2. If we find such a group, we insert it into L that corresponds to a set of QI-groups to be output. Finally, we continue this process for each remaining tuple in the table iteratively until all tuples in the table have been processed.

- **Phase 2 (Remaining Assignment Phase):** In this phase, we process each tuple in T_r. For each tuple t in T_r, we do the following. Initially, t is inserted into the group that has the smallest number of tuples that are incompatible with t. Then, we iteratively merge this group with another group such that the merged group has the greatest number of sensitive values until t has at least l valid sensitive values in the merged group

The detailed algorithm can be found in Algorithm 7.

Table 3.20: A table anonymized from Table 3.18 by Injector

Sex	Zipcode	GID
male	54321	G_1
female	54323	G_2
male	54322	G_1
male	54327	G_4
female	32134	G_2
female	32135	G_3
female	32138	G_3
female	32139	G_4
...	...	...

(a) QI Table

GID	Disease
G_1	Heart Disease
G_2	Ovarian Cancer
G_1	Flu
G_4	Stomach Virus
G_2	HIV
G_3	Ovarian Cancer
G_3	flu
G_4	fever
...	...

(b) Sensitive table

Let us illustrate with the example we have been considering. For Step 1, we assume that we use some existing data mining algorithms to find the negative association rule "Male $\rightarrow \neg$ Ovarian Cancer".

For Step 2, from Table 3.18, we first find Mary's tuple. Then, we form a QI-group consisting of {Mary}. Then, we find a tuple that is compatible with Mary's tuple. There are many tuples that are compatible with Mary's tuple (e.g., Emily's and Helen's tuples). Let us assume that we choose Emily's tuple. We form a group {Mary, Emily}. In this group, each tuple has 2 valid sensitive tuples. Similarly, we perform this process and we obtain the table as shown in Table 3.20.

3.6. KNOWLEDGE MINED FROM THE MICRODATA

Algorithm 7 Algorithm Injector

Require: table T, privacy requirement $\mathcal{R}$ of l-diversity
Ensure: table T^* that satisfies l-diversity
1: $L \leftarrow \emptyset$
2: // Phase 1 (Initial Grouping Phase)
3: **while** $|T| \geq l$ **do**
4:    select the tuple t that has the largest number of incompatible tuples
5:    remove t from T
6:    $G \leftarrow \{t\}$
7:    **while** $|G| < l$ **do**
8:      find a tuple t' such that t' is compatible with G and the new group (formed by inserting t' into G) has the smallest number of incompatible tuples
9:      **if** such t' exists **then**
10:         $T \leftarrow T - \{t'\}$
11:         $G \leftarrow G \cup \{t'\}$
12:      **else**
13:         insert each tuple t'' in $G/\{t\}$ into T
14:         insert t into T_r
15:      **end if**
16:    **end while**
17:    insert G into L
18: **end while**
19: insert all tuples in T into T_r
20: // Phase 2 (Remaining Assignment Phase)
21: **for** each tuple t in T_r **do**
22:    find G in L that has the smallest number of tuples that are incompatible with t
23:    insert t into G
24:    **while** t has less than l valid sensitive values in G **do**
25:      select G' in L such that $G' \cup G$ has the greatest number of sensitive values
26:      $G \leftarrow G \cup G'$
27:      remove G' from L
28:    **end while**
29: **end for**
30: **return** T^*

3.7 KNOWLEDGE MINED FROM THE PUBLISHED DATA

The previous section demonstrated that knowledge can be mined from the microdata and can be used by an adversary. In fact, knowledge can also be mined from the *published data* or the *anonymized data* to compromise individual privacy [Kifer, 2009; Wong et al., 2009b]. One of the main purposes of data publishing is to allow analysis through data mining by which patterns can be discovered from the published data. The knowledge derived from the published data can be used by an adversary to deduce some sensitive information.

Let us illustrate the problem with an example. Suppose Table 3.21 is to be anonymized for publication in which one of the QI attributes is `Nationality` and the sensitive attribute is `Disease`. Note that there can be other QI attributes in this table such as sex and zip code. For the sake of illustration, we list `Nationality` only. As before, assume that each tuple in the table is owned by an individual and each individual owns at most one tuple.

Suppose that the data owner wants to anonymize this table and publishes the anonymized dataset given in Table 3.22 that is obtained by bucketization satisfying 2-diversity. The intention is that each individual cannot be linked to a disease with a probability of more than 0.5. However, *does this table protect individual privacy sufficiently?*

Table 3.21: An example

Name	Nationality	...	Disease
Alex	American	...	Heart Disease
Tanaka	Japanese	...	Flu
Aoi	Japanese	...	Flu
Hoshi	Japanese	...	Stomach Virus
Emily	French	...	HIV
Mika	Japanese	...	Diabetes
...	...	...	...

Table 3.22: A 2-diverse dataset anonymized from Table 3.21

Nationality	...	GID
American	...	G_1
Japanese	...	G_1
Japanese	...	G_2
Japanese	...	G_2
French	...	G_3
Japanese	...	G_3
...	...	...

(a) QI Table

GID	Disease
G_1	Heart Disease
G_1	Flu
G_2	Flu
G_2	Stomach Virus
G_3	HIV
G_3	Diabetes
...	...

(b) Sensitive table

Let us examine the QI-group with GID equal to G_1 as shown in Table 3.22. We also refer to the QI-group by G_i. In G_1, Heart Disease and Flu are values of the sensitive attribute Disease. It *seems* that each of the two individuals, Alex and Tanaka, in this group has a 50% probability of linking to Heart Disease and 50% probability of linking to Flu. The reason why the probability is interpreted as 50% is that the analysis is based on this group *locally* without any additional information.

However, from the *entire published table* containing *multiple* groups, the adversary may discover some interesting patterns *globally*. For example, suppose the published table consists of many QI-groups like G_2 with all Japanese with no occurrence of Heart Disease. At the same time, there are many QI-groups like G_3 containing some Japanese without Heart Disease. The pattern that Japanese rarely suffer from Heart Disease can be uncovered. Note that it is very likely that such an anonymized data may be published by conventional anonymization methods, given the fact that Heart Disease occurs rarely among Japanese. With the pattern uncovered, the adversary can say that Tanaka, being a Japanese, has less chance of having Heart Disease. S/he can deduce that Alex, being an American, has a higher chance of having Heart Disease. The intended 50% threshold is thus violated.

The anonymized data can be seen as *imprecise* or *uncertain data* [Burdick et al., 2005, 2007], and an adversary can uncover interesting patterns since the published data must maintain high data utility [Wong et al., 2007; Xiao and Tao, 2006b; Zhang et al., 2007]. Since it is easy to obtain the knowledge mined from the anonymized dataset, all existing privacy models suffer from privacy breaches.

In Table 3.22, there are only two *local possible worlds* for assigning the disease values to the two individuals in G_1: (1) w_1 : Alex is linked to Heart Disease and Tanaka is linked to Flu and (2) w_2 : Alex is linked to Flu and Tanaka is linked to Heart Disease. To construct a probability distribution over the domain of the real world, the simplest definition is based on the *random world assumption* that *all the possible worlds are equally likely*, or *each world has the same probability*.

If we publish group G_1 alone, the random world assumption is a good principle in the absence of other information. However, when many groups are published together as typically the case, the groups with Japanese in contrast to other groups contribute to a statement that their members are not likely linked to Heart Disease. This statement means that the *probability* (or *weight*) of the possible world w_1 is much greater than that of w_2.

Most works such as *l*-diversity [Machanavajjhala et al., 2006], *t*-closeness [Li and Li, 2007], (k, e)-anonymity [Zhang et al., 2007] and *m*-confidentiality [Wong et al., 2007] adopt the random world assumption. The source of attack of the adversary with the knowledge mined from the published data is to apply the more complete model of the *weighted possible worlds* in which different possible worlds have different probabilities. The probability of each possible world can be derived from the published table by some data analysis techniques and some data mining techniques [Kifer, 2009; Wong et al., 2009b].

Although Kifer [2009] and Wong et al. [2009b] discover the possibility of privacy breaches according to the knowledge mined from the published data, there are still no existing works that

3. ONE-TIME DATA PUBLISHING

study how a table can be generated that is resistant to this kind of privacy breaches. It is interesting to study this open problem in the future.

3.8 HOW TO USE PUBLISHED DATA

In our previous discussion, we focused on describing how to anonymize the data and how to protect individual privacy. However, when the data are anonymized or modified, how do users (or data analysts/data miners) analyze the published data that are different from the original?

Table 3.23: A microdata

Gender	Nationality	Age	Disease
male	Japanese	26	HIV
female	Malaysian	30	flu
female	American	36	HIV
female	Canadian	40	flu
male	American	40	HIV
male	Chinese	36	flu
male	Canadian	70	fever
male	Chinese	76	Heart Disease

Table 3.24: An example illustrating how to use the published data

Gender	Nationality	Age	GID
male	Japanese	26	1
female	Malaysian	30	1
female	American	36	2
female	Canadian	40	2
male	American	40	3
male	Chinese	36	3
male	American	70	4
male	Chinese	76	4

(a) QI table

GID	Disease
1	HIV
1	flu
2	HIV
2	flu
3	HIV
3	flu
4	fever
4	Heart Disease

(b) Sensitive table

Since the anonymized data are different from the original data, the method of analyzing the anonymized data is also different. Consider the microdata as shown in Table 3.23. If we are interested in knowing the total number of male individuals suffering from HIV, from this table, we can obtain the number equal to 2. However, the task of obtaining this number is not trivial when the data are anonymized. Suppose the anonymized data are generated by bucketization as shown in Table 3.24. In the anonymized data, there are four QI-groups. The first QI-group and the last two QI-groups contain some male individuals, but the second QI-group does not. Thus, we need to focus on analyzing the first QI-group and the last two QI-groups. Consider the last QI-group containing all male individuals. Since it does not contain any HIV value, we are sure that there is no male individual suffering from HIV in this group. Consider the second-to-last QI-group containing all male individuals. This group contains one HIV value. We are sure that there is only one male individual suffering from HIV in this QI-group. Consider the first QI-group containing one male individual and one female individual. In this group, there is one HIV value. Note that, according to all the QI-groups, we know that each individual has 50% probability of linking to HIV. Thus, we estimate that the expected number of male individuals suffering from HIV is 0.5 in this group. Thus, according to the QI-groups containing male individuals, the total expected number of male individuals suffering from HIV is equal to 0 + 1 + 0.5 = 1.5.

From this example, we can see that the analysis over the anonymized data is different from that over the original data. In this section, we will describe some methods for analyzing the anonymized data.

3.8.1 AGGREGATE QUERIES

In this section, we describe how the *data analyst* uses aggregate queries for data analysis. This technique is commonly adopted in an anonymized table generated by bucketization [Li and Li, 2008; Wong et al., 2007; Xiao and Tao, 2006b]. Using aggregate queries, the *data owner* can compare the quality of the anonymized data with that of the original microdata by measuring the relative error ratio in answering an aggregate query.

Specifically, the form of the aggregate query is as follows:

```
SELECT COUNT(*)
FROM Unknown-Microdata
WHERE pred(A_1^{qi}) AND ... AND pred(A_{qd}^{qi}) AND pred(A^s)
```

In the above query, qd denotes the query dimensionality, A_j^{qi} denotes the j-th attribute to be queried and A^s denotes the sensitive attribute. For any attribute A, the predicate $pred(A)$ has the form $(A = x_1 \text{ OR } A = x_2 \text{ OR } ... \text{ OR } A = x_b)$ where x_i is a random value in the domain of A, for $1 \leq i \leq b$. The value of b depends on the *expected query selectivity s*

$$b = \lceil |A| \cdot s^{1/(qd+1)} \rceil$$

where $|A|$ is the domain size of A. If the value of s is set higher, the number of selection conditions in $pred(A)$ will be greater. The aggregate query is performed with the original data set and the anonymized data set, Unknown-Microdata would be replaced with the respective data set. Let the result obtained from the original data set be act, and the result from the anonymized data set be est.

In the example illustrated at the beginning of this chapter, recall that we want to know the total number of male individuals suffering from HIV. Since we would like to know about male individuals, the QI attribute that we are interested in is Gender. Thus, the total number of query

QI attributes, qd, is 1. Let A be attribute Gender. Since there are two possible values in Gender, $|A| = 2$. Suppose s is set to 0.05. Then, for attribute Gender,

$$\begin{aligned} b &= \lceil |A| \cdot s^{1/(qd+1)} \rceil \\ &= \lceil 2 \cdot 0.05^{1/(1+1)} \rceil \\ &= 1 \end{aligned}$$

For attribute Gender, the predicate $pred(A)$ has the form $(A = x_1)$. Suppose x_1 is male. In our example, A_s is attribute Disease with three possible values. Similar to attribute A, we obtain $b = 1$ for attribute A_s. Thus, the query becomes

```
SELECT COUNT(*)
FROM Unknown-Microdata
WHERE Gender = Male AND Disease = HIV
```

Since the anonymized data set contains two tables, namely the QI table and the sensitive table, we estimate the result as follows. When a tuple in the QI table with GID = id is matched with $pred(A_1^{qi}),..., pred(A_{qd}^{qi})$, the count of this tuple that contributes to the final result is estimated as the fraction of tuples that are matched with $pred(A^s)$ over the tuples with GID = id in the sensitive table. est is the sum of all estimated values of the matched tuples.

$$\text{relative error ratio} = \frac{|act - est|}{act}$$

In the experiments conducted by Xiao and Tao [2006b], by default, s is set to 0.05 and qd is set to the QI size. Besides, for each evaluation, 10,000 random queries are generated and executed, and the average relative error ratio is computed.

In our example, since we obtain a count over the original table equal to 2 and a count over the anonymized table equal to 1.5, we have $act = 2$ and $est = 1.5$. Thus,

$$\begin{aligned} \text{relative error ratio} &= \frac{|act - est|}{act} \\ &= \frac{|2 - 1.5|}{2} \\ &= 0.25 \end{aligned}$$

3.8.2 INFORMATION LOSS

In the previous section, we discussed how to analyze anonymized data generated by bucketization. In this section, we describe how to analyze the anonymized data generated by generalization (either local recoding or global recoding). When a table T is anonymized to a more generalized table T^*, it is of interest to measure the information loss that is incurred. In Section 2.2, we described one metric called distortion ratio to measure the information loss of the anonymized table. This metric is based on the *height* of generalized values in the taxonomy. In the following, we give another metric [Wong et al., 2007; Xiao and Tao, 2006a] to measure the information loss that is based on the *coverage* of generalized values (or more specifically, the total number of descendant leaf nodes of generalized values in the taxonomy.)

Definition 3.4 Coverage and Base. Let $\mathcal{T}$ be the taxonomy of a QI attribute. The *coverage* of a generalized QI value v^*, denoted by $coverage[v^*]$, is given by the number of the descendant leaf values of value v^* in $\mathcal{T}$. The *base* of the taxonomy $\mathcal{T}$, denoted by $base(\mathcal{T})$, is the number of leaf values in the taxonomy.

For example, in Figure 3.2, $base(\mathcal{T}) = 9$, and $coverage[university] = 2$ since undergrad and postgrad can be generalized to university.

A weight can be assigned for each attribute A, denoted by $weight(A)$, to reflect the users' opinion regarding the significance of information loss in different attributes. Let $t.A$ denote the value of A in tuple t.

In the following, we give the definition of the information loss of the published table T^* based on the coverage of values. Intuitively, if T^* contains more values with large coverage, then the information loss of T^* is larger.

Definition 3.5 Information Loss. Let table T^* be an anonymization of table T obtained by means of a mapping function f. Let $\mathcal{T}_A$ be the taxonomy for attribute A that is used in the mapping and v_A^* be the nearest common ancestor of $t.A$ and $f(t).A$ in $\mathcal{T}_A$ where t is a tuple in T. The *information loss* of a tuple t^* in T^* introduced by f is given by

$$\mathcal{IL}(t^*) = \sum_{A \in QI} \{\mathcal{IL}(t^*, A) \times weight(A)\}$$

where

$$\mathcal{IL}(t^*, A) = \begin{cases} \frac{coverage[v_A^*]-1}{base(\mathcal{T}_A)-1} & \text{if } base(\mathcal{T}_A) > 1 \\ 0 & \text{if } base(\mathcal{T}_A) = 1 \end{cases}$$

The *information loss* of the published table T^* is given by

$$\mathcal{IL}(T^*) = \frac{\sum_{t^* \in T^*} \mathcal{IL}(t^*)}{|T^*|}$$

If $f(t).A = t.A$, then $f(t).A$ is a leaf value, the nearest common ancestor $v_A^* = t.A$, and $coverage[v_A^*] = 1$. If this is true for all A's in QI, then $\mathcal{IL}(t^*) = 0$, which means there is no information loss. If $t.A$ is generalized to the root of taxonomy $\mathcal{T}_A$, then the nearest common ancestor v_A^* = the root of $\mathcal{T}_A$. Thus, $coverage[v_A^*] = base(\mathcal{T}_A)$ and, if this is the case for all A's in QI, then $\mathcal{IL}(t^*) = 1$.

The above definition of information loss is based on the *content* of each *generalized* tuple. There are other definitions of information loss related to the *size* of QI groups and the total *number* of QI groups. Some definitions are shown as follows.

- The *average QI-group size* is considered as a measure for information loss [Machanavajjhala et al., 2006]. Intuitively, if the average QI-group size is larger, there will be more information loss.

- The discernability model proposed by Bayardo and Agrawal [2005] assigns a penalty to each tuple t as determined by the square of the *size* of the QI-group for t. The total penalty of all tuples corresponds to the information loss.

- The *normalized average QI-group size* can be used as a metric [LeFevre et al., 2006], which is given by the total number of tuples in the table divided by the product of the total number of QI-groups and a value k (for k-anonymity). Here, the best case occurs when each QI-group has size k.

These definitions related to the size of QI groups are simple to compute. However, unlike the information loss defined in Definition 3.5, they cannot distinguish a QI-group with another QI-group of the same size even though the tuples in one group and the tuples in the other group are very different. For example, consider a QI-group G_1 contains two tuples that both correspond to Japanese, and another QI-group G_2 contains one tuple that corresponds to Canadian and another tuple that corresponds to American. If the generalization is used for anonymization, in the anonymized table, G_1 contains two tuples for Japanese and G_2 contains two tuples for North American. Even though G_1 and G_2 have the same size (which are considered as the same information loss in the above definitions), obviously, G_1 contains more detailed information compared with G_2 because there are generalized values in G_2 but not in G_1.

3.8.3 EVALUATION WITH DATA MINING AND DATA ANALYSIS TOOLS

In the previous sections, we *evaluate* how "good" the anonymized data is by the information loss of the published table. Alternatively, we can also evaluate the anonymized data with some existing data mining and data analysis tools.

Suppose that the data is anonymized by global recoding. Thus, it contains consistent values, and we can make use of some existing data mining and data analysis tools to evaluate how "good" the anonymized data is. Some examples of data mining and data analysis used in the literature are [Fung et al., 2005; Lefevre et al., 2008]:

- Classification/Regression Task: C4.5 classifier and a Naive Bayesian classifier can be used to evaluate the classification accuracy over the anonymized data. Fung et al. [2005] indicate that the classification accuracy of Naive Bayesian over the anonymized table that satisfies k-anonymity is higher than that over the original table when $k < 200$. This result suggests that the original table contains a lot of specific information that can be regarded as noise in the classification process. However, the anonymized table usually generalizes the data and thus the specific information is removed. Thus, the accuracy of the anonymized table is higher compared with the original table.

- Data Analysis Task: Some SQL queries with a set of selection predicates and some aggregate functions such as SUM, MIN and AVG can be used to evaluate the anonymized table.

3.8.4 QUERYING OVER AN UNCERTAIN DATABASE

The anonymized table can be regarded as *uncertain data* or *probabilistic data*. The query operators over uncertain data can be applied on the analysis over the anonymized data.

For the sake of illustration, let us use the anonymized generated by bucketization as shown in Table 3.25. Similar arguments can be made for a table generated by generalization.

Table 3.25: An example illustrating how to use the published data

Gender	Nationality	Age	GID
male	Japanese	26	1
female	Malaysian	30	1
female	American	36	2
female	Canadian	40	2

(a) QI table

GID	Disease
1	HIV
1	flu
2	HIV
2	flu

(b) Sensitive table

The major idea of analyzing uncertain data is described as follows. We first list all *possible* original microdata that can generate the table like Table 3.25. Each possible original microdata is called a *possible world*. Consider the first QI-group. There are two possible worlds for this *QI-group*. The first possible world is that the male individual is linked to HIV and the female individual is linked to flu. The second possible world is that the male individual is linked to flu and the female individual is linked to HIV. Similarly, there are two possible worlds for the second *QI-group*. Table 3.26 shows all the four possible worlds, namely w_1, w_2, w_3 and w_4, for the *whole table*, namely Table 3.25.

After we enumerate all possible worlds, we can perform any traditional queries, including aggregate queries (discussed in Section 3.8.1) and data mining tools (discussed in Section 3.8.3), over each possible table and obtain the correspondence results.

If the result is a number like the total number of male individuals suffering from HIV, we can take the average value obtained from the queries performed over all the possible worlds. For example, consider the aggregate query for the total number of male individuals suffering from HIV. We obtain the answers from w_1, w_2, w_3 and w_4 equal to 1, 1, 0 and 0, respectively. Assume that each possible world occurs with equal probability. Then, the expected total number of male individuals suffering from HIV is equal to $(1 + 1 + 0 + 0)/4 = 0.5$.

If the result is not a single number (e.g., a set of tuples), then the analysis will become complicated. However, some existing works about uncertain databases study this complicated case.

Table 3.26: Four possible worlds of Table 3.25

Gender	Nationality	Age	Disease
male	Japanese	26	HIV
female	Malaysian	30	flu
female	American	36	HIV
female	Canadian	40	flu

(a) Possible world w_1

Gender	Nationality	Age	Disease
male	Japanese	26	HIV
female	Malaysian	30	flu
female	American	36	flu
female	Canadian	40	HIV

(b) Possible world w_2

Gender	Nationality	Age	Disease
male	Japanese	26	flu
female	Malaysian	30	HIV
female	American	36	HIV
female	Canadian	40	flu

(c) Possible world w_3

Gender	Nationality	Age	Disease
male	Japanese	26	flu
female	Malaysian	30	HIV
female	American	36	flu
female	Canadian	40	HIV

(d) Possible world w_4

Details can be found in the existing works about uncertain databases [Aggarwal and Yu, 2009; Cormode et al., 2009; Hua et al., 2008; Soliman et al., 2007].

3.9 CONCLUSION

In this chapter, we learnt that the adversary can have different kinds of background knowledge to breach individual privacy. Some examples of background knowledge are:

- the knowledge about the distribution of sensitive values (Section 3.2)
- the knowledge about the linkage of individuals to sensitive values (Section 3.3)
- the knowledge about the relationship among individuals (Section 3.4)
- the knowledge about anonymization (Section 3.5)

- the knowledge mined from the microdata (Section 3.6)
- the knowledge mined from the published data (Section 3.7)

According to different kinds of background knowledge, the adversary can breach individual privacy. Thus, the data publisher should modify the data such that the modified data is resistant to different kinds of background knowledge.

In this chapter, we also studied some metrics to measure the information loss of the modified data and some methods to analyze the published data (Section 3.8).

CHAPTER 4
Multiple-Time Data Publishing

In the previous chapter, we studied how to anonymize data when they are published only once. However, in practice, a data owner typically publishes data repeatedly over time. One straightforward method to address this situation is to anonymize the data as before each time it is published such that each version of the data satisfies a privacy requirement. However, the collection of all published data may still allow an adversary to breach privacy because there can be some correlations among the versions of the published data.

In this chapter, we study how to publish the data taking into consideration the correlations among versions of published data. There are at least two types of correlations to consider in multiple-time data publishing.

- *Individual-based correlation:* This happens when the information about a single individual appears in multiple published tables.

- *Sensitive value-based correlation:* This occurs when the information about the linkage of a sensitive value to an individual remains unchanged with high probability when this individual appears in later published tables. Note that individual-based correlation does not consider any sensitive values, but sensitive value-based correlation does.

In Section 4.1, we first describe how the adversary can make use of individual-based correlation to breach privacy and give a method to prevent this attack. In Section 4.2, we consider the sensitive value-based correlation and describe a method to deal with it.

4.1 INDIVIDUAL-BASED CORRELATION

Some microdata change over time but others remain unchanged. The former is called *dynamic microdata* while the latter is called *static microdata*. Section 4.1.1 describes data publishing from static microdata. and Section 4.1.2 describes data publishing from dynamic microdata.

4.1.1 DATA PUBLISHING FROM STATIC MICRODATA

Wang and Fung [2008] propose a privacy model that publishes different *views* over a static table (Figure 4.1). Given a static table with many attributes, according to some users' request, the data owner projects a set of chosen attributes over the static table and generates a view of the original table. Then, it publishes this view. At a later time, according to the request of other users, it projects another set of chosen attributes over the static table, and it generates another view and publishes it. Since

4. MULTIPLE-TIME DATA PUBLISHING

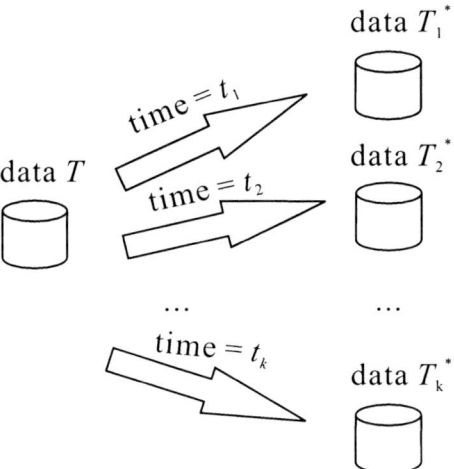

Figure 4.1: Data publishing from static microdata

two different views have been published and the correlation among individuals is not considered, individual privacy may be breached.

Consider that the data owner has the microdata as shown in Table 4.1(a). The data owner has the following privacy requirement:

"With high probability, the combination of attributes Gender and Age should not be linked to Disease."

Let v be a value pair representing a possible value of the combination of values of Gender and Age. For example, v is equal to (*Male*, 80). Let $n(v)$ denote the number of tuples containing value v and $n(v, s)$ denote the number of tuples containing value v and the sensitive value s. Formally, the privacy requirement is described as follows: given a parameter $\alpha \in [0, 1]$, for each possible value v and each possible sensitive value s, the following inequality holds:

$$\frac{n(v, s)}{n(v)} \leq \alpha$$

Note that the attributes specified in this privacy requirement involves three attributes only, Gender, Age and Disease. Attributes Gender and Age are considered quasi-identifiers, and attribute Disease is considered as a sensitive attribute. In this example, attribute Disease contains one sensitive value, namely HIV, and two non-sensitive values, namely flu and fever.

This privacy requirement is called (X, Y)-*privacy* where X and Y are two sets of attributes in the microdata and $X \cap Y = \emptyset$. In our example, X is equal to {Gender, Age} and Y is equal to {Disease}.

Suppose that we set $\alpha = 0.5$. Consider that we want to obtain the microdata to study the relationship among attributes Education, Age and Disease. Since the set of these three attributes does not contain all attributes specified in the privacy requirement, a view involving these attributes can be generated and published (Table 4.1(b)). After a period of time, some other request may be made for the data related to attributes Gender and Education. Again, since the requested attributes are not related to all attributes specified in the privacy requirement, a view over these attributes may be generalized and released (Table 4.1(c)). However, the release of the second table may be problematic. An adversary that has access to both released tables may perform a *join* operation over them via attribute Education. The joined table is shown in Table 4.2. Suppose the adversary knows that the QI attributes (Gender, Age) of Peter have values (Male, 80). According to the joined table, the adversary deduces that Peter suffers from HIV.

Table 4.1: Different views of the static table

Gender	Education	Age	Disease
Male	Elementary	80	HIV
Male	Secondary	60	Fever
Female	Secondary	36	Flu
Female	University	40	Flu

(a) A microdata

Education	Age	Disease
Elementary	80	HIV
Secondary	60	Fever
Secondary	36	Flu
University	40	Flu

(b) A table projected on attributes Education, Age and Disease

Gender	Education
Male	Elementary
Male	Secondary
Female	Secondary
Female	University

(c) A table projected on attributes Gender and Education

In order to avoid this privacy breach, Wang and Fung [2008] propose a privacy model that protects individual privacy according to the privacy requirement by adopting the concept of *lossy join*. Consider microdata T. Let $T_1^*, T_2^*, ..., T_{h-1}^*$ be the previously published data. Consider that the data owner wants to publish a table T_h that is projected on a set of chosen attributes over table T. The data owner wants to modify table T_h to table T_h^* such that the join of all previously published tables and table T_h^* satisfies the privacy requirement (i.e., (X, Y)-privacy). We adopt one of the well-known algorithms mentioned in Section 2.3 (e.g., top-down approach) to generalize T_h to T_h^* in order to avoid privacy breaches.

Table 4.2: A joined table between Table 4.1(b) and Table 4.1(c) via attribute Education

Gender	Education	Age	Disease
Male	Elementary	80	HIV
Male	Secondary	60	Fever
Female	Secondary	36	Cancer
Female	Secondary	60	Fever
Male	Secondary	36	Cancer
Female	University	40	Flu

Consider our running example. Originally, T_1^* (Table 4.1(b)) that was projected on Education and Age was published. Now, we want to publish table T_2 (Table 4.1(c)) that is projected on Gender and Education, and generalize it to T_2^*. Consider that we use the top-down approach mentioned in Section 2.3. The first step is to fully generalize each attribute of T_2 such that each value becomes * as shown in Table 4.3(a). We can try to perform a join operation over Table 4.1(b) and Table 4.3(a) and obtain the joined table as shown in Table 4.3(b). Note that the domain of attribute Education in Table 4.1(b) contains some *specific* values like Elementary, Secondary and University, and the domain of attribute Education in Table 4.3(a) contains some *generalized* values like *. In order to perform the join operation, value * can be considered a special value that can be Elementary, Secondary or University. Specifically, the first tuple in Table 4.1(b) can be joined with each of the four tuples in Table 4.3(a). Similar arguments can be made on the second, the third and the fourth tuples in Table 4.1(b). Furthermore, if the adversary knows that the QI attributes (Gender, Age) of Peter have values (Male, 80), according to this joined table (Table 4.3), s/he can also deduce that Peter suffers from HIV.

It is easy to verify that the joined result between one of possible generalizations of T_2 (Table 4.1(c)) and T_1^* (Table 4.1(b)) will also give the adversary opportunities to breach individual privacy. This means that the first release T_1^* discloses too much information.

Suppose that the first release T_1^* is Table 4.4(a) instead of Table 4.1(b). Consider that we want to publish T_2 (Table 4.1(c)). If we adopt the top-down approach to generalize T_2, we publish T_2^* as shown in Table 4.4(b). Then, the join result between Table 4.4(a) and Table 4.4(b) is shown in Table 4.5. It is easy to see that each possible value of the combination of Gender and Age is not linked to any sensitive value with more than 50% probability. For example, value (Person, [60-80]) is linked to one occurrence of HIV (a sensitive value) and two occurrences of flu (a non-sensitive value). The probability that value (Person, [16-80]) is linked to HIV is 1/3 (that is smaller than 1/2). Thus, the published table T_1^* (Table 4.4(a)) and the published table T_2^* (Table 4.4(b)) protect individual privacy.

Table 4.3: Different views of the static table

Gender	Education
*	*
*	*
*	*
*	*

(a) A fully generalized table of T_2 that is projected on attribute Gender and attribute Education

Gender	Education	Age	Disease
*	Elementary	80	HIV
*	Elementary	80	HIV
*	Elementary	80	HIV
*	Elementary	80	HIV
*	Secondary	60	Fever
*	Secondary	60	Fever
*	Secondary	60	Fever
*	Secondary	60	Fever
*	Secondary	36	Cancer
*	Secondary	36	Cancer
*	Secondary	36	Cancer
*	Secondary	36	Cancer
*	University	40	Flu
*	University	40	Flu
*	University	40	Flu
*	University	40	Flu

(b) A joined table between Table 4.1(b) and Table 4.3(a)

4.1.2 DATA PUBLISHING FROM DYNAMIC MICRODATA

In the previous section, we described a privacy model that publishes different views from a static table. However, in practice, the data kept by a data owner is likely to change over time, as shown in Figure 4.2 where T_1 is updated to T_2. There are three possible cases of data change over time:

- *Insertion:* Some tuples that did not exist in some previous published tables are inserted in some later published tables;

- *Deletion:* Some tuples that existed in some previous published tables are deleted in some later published tables;

- *Update:* The attribute values of some tuples are updated.

Table 4.4: The published views that avoid privacy breaches

Education	Age	Disease
Elementary	[60-80]	HIV
Secondary	[60-80]	Fever
Secondary	[36-40]	Cancer
University	[36-40]	Flu

(a) A table projected on attributes Education, Age and Disease (T_1^*)

Gender	Education
Person	Elementary
Person	Secondary
Person	Secondary
Person	University

(b) A table projected on attributes Gender and Education (T_2^*)

Table 4.5: A joined table between Table 4.4(b) and Table 4.4(c) via attribute Education

Gender	Education	Age	Disease
Person	Elementary	[60-80]	HIV
Person	Secondary	[60-80]	Fever
Person	Secondary	[36-40]	Flu
Person	Secondary	[60-80]	Fever
Person	Secondary	[36-40]	Flu
Person	University	[36-40]	Flu

In this section, we study the privacy model called *m-invariance* [Xiao and Tao, 2007] that considers privacy protection with the first two cases *only* (i.e., insertions and deletions). Note that, interestingly, the third case (i.e., update) *cannot* be handled sufficiently by a privacy model that only considers the first two cases (although an update operation can simulated by first executing the deletion operation and then executing the insertion operation). We will explain in detail in Section 4.2 how individual privacy can be protected in all three cases.

Let us illustrate the issue with an example. Table 4.6(a) (T_1) and Table 4.6(b) (T_2) show the microdata at times t_1 and t_2. Consider that the data owner wants to publish tables such that the probability that an individual is linked to any sensitive value is at most a given threshold. Suppose that the data owner uses l-diversity as the privacy requirement. The given threshold in the above privacy requirement is set to $1/l$. Note that table T_2 is different from table T_1 since there are three tuples removed from T_1 (namely, Bob's, Fred's and Iris's tuples in T_1) and one new tuple inserted (namely, Gary's tuple).

4.1. INDIVIDUAL-BASED CORRELATION

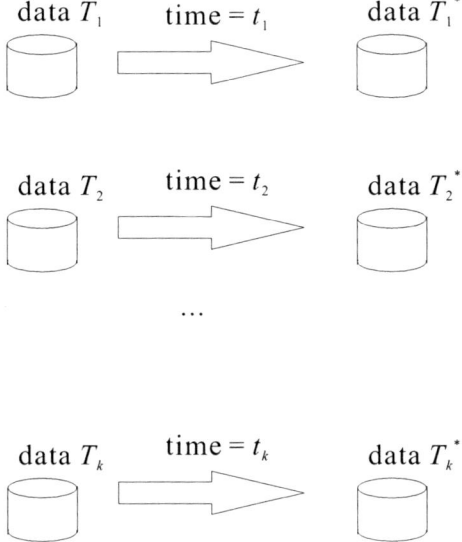

Figure 4.2: Data publishing from dynamic microdata

If the data owner publishes these two tables separately without considering the correlation between these two tables, the two tables will be published as shown in Table 4.7(a) and Table 4.7(b). Note that each of the published tables satisfies 2-diversity; thus, in each published table, every individual is linked to a sensitive value with probability 0.5. It seems that individual privacy can be protected.

However, if the correlation between the two published tables is considered, an adversary can breach privacy. Suppose that the adversary knows Alice's QI attributes. S/he can deduce that Alice is in the first QI-group in Table 4.7(a). Based on this table, s/he is sure that Alice is linked to either HIV or fever. Similarly, according to Table 4.7(b), since Alice is in the first QI-group, s/he can deduce that Alice is linked to either HIV or Cancer. By combining these two pieces of information, s/he is sure that Alice is linked to HIV only. Thus, Alice's privacy is breached.

The major reason why the adversary can breach individual privacy is that the QI-group originally containing Alice's and Bob's tuples in T_1^* has two values for the sensitive attribute, HIV and fever, and this QI-group containing Alice's tuple in T_2^* changes and contains two values in the sensitive attribute, HIV and Cancer. The absence of value fever in the new QI-group provides a crucial piece of information to the adversary allowing the breach of Alice's privacy.

We can see that, no matter how we anonymize Table 4.6(b), Alice's privacy is breached. This is because there is no other tuple with value fever in Table 4.6(b). Whenever the new QI-group containing Alice's tuple does not contain fever, Alice's privacy is breached.

Motivated by this observation, Xiao and Tao [2007] propose to introduce some counterfeit tuples in the published table. Suppose that Table 4.6(a) was published as Table 4.7(a). The data owner

Table 4.6: A microdata for illustrating m-invariance

Name	Age	Zipcode	Disease
Alice	25	54321	HIV
Bob	26	54320	fever
Clement	32	12350	cancer
David	34	12352	flu
Emily	47	35214	Heart Disease
Fred	49	35215	flu
Helen	22	54200	Cancer
Iris	26	54211	flu

(a) T_1

Name	Age	Zipcode	Disease
Alice	25	54321	HIV
Clement	32	12350	cancer
David	34	12352	flu
Emily	47	35214	Heart Disease
Helen	22	54200	Cancer
Gary	49	35214	flu

(b) T_2

publishes the second table as shown in Table 4.8(a) with two counterfeit tuples, namely c_1 and c_2. We can see that the QI-group containing Alice's tuple contains a counterfeit tuple represented by c_1. Note that this QI-group in Table 4.8(a) contains the same set of values for Disease as the QI-group containing Alice's tuple in Table 4.7(a) (i.e., HIV and fever).

From the perspective of an adversary, a counterfeit tuple is indistinguishable from the real tuple in the original table. Thus, according to Table 4.7(a) and Table 4.8(a), the adversary may deduce that Alice is linked to HIV with a probability at most 0.5.

The key idea of this technique is to maintain the "invariance" of the values in the sensitive attribute in each QI-group over different timestamps. Specifically, each individual is assigned a *signature* in each published table T_i^*. The signature of an individual in table T_i^* is defined to be a set of values in the sensitive attribute in the QI-group containing this individual's tuple. For example, Alice's signature in T_1^* is equal to {HIV, fever}, which means that Alice is linked either HIV or fever in the original microdata. A privacy model called m-invariance is defined such that the signature of each individual does not change over time. Thus, in the published tables, each individual is guaranteed to be linked to any sensitive value with probability at most $1/m$. Since Alice's signature in T_1^* is equal to {HIV, fever}, m-invariance requires that Alice's signature must be equal to {HIV, fever} in each of the later releases. Specifically, Alice's signature in T_2^* (Table 4.8(a)) is also equal to {HIV, fever}.

Table 4.7: Published tables of Tables 4.6(a) and (b)

Name	GID	Age	Zipcode	Disease
Alice	1	[25-26]	[54320-54321]	HIV
Bob	1	[25-26]	[54320-54321]	fever
Clement	2	[32-34]	[12350-12352]	cancer
David	2	[32-34]	[12350-12352]	flu
Emily	3	[47-49]	[35214-35215]	Heart Disease
Fred	3	[47-49]	[35214-35215]	flu
Helen	4	[22-26]	[54200-54211]	Cancer
Iris	4	[22-26]	[54200-54211]	flu

(a) T_1^* satisfying 2-diversity

Name	GID	Age	Zipcode	Disease
Alice	1	[22-25]	[54200-54321]	HIV
Helen	1	[22-25]	[54200-54321]	Cancer
Clement	2	[32-34]	[12350-12352]	cancer
David	2	[32-34]	[12350-12352]	flu
Emily	3	[47-49]	[35214-35214]	Heart Disease
Gary	3	[47-49]	[35214-35214]	flu

(b) T_2^* satisfying 2-diversity

Although T_2^* (Table 4.8(a)) alone can guarantee the privacy protection, in order to improve the utility of the published table T_2^*, a table (Table 4.8(a)) storing the number of counterfeit tuples in each QI-group of the generalized table is published in addition to T_2^*. This table is called a *counterfeit table*. Recall that the objective is to publish a table that satisfies some privacy requirements and achieves high utility. With the counterfeit table, the results of some types of data analysis over the anonymized table can be improved. For example, as discussed in Section 3.8.1, the accuracy of aggregate queries can be used to measure the utility of an anonymized table. The counterfeit table reduces the error of aggregate queries. Suppose that we are interested in knowing the total number of individuals who are at most 35 years old. According to Table 4.8(a) *alone*, we find 6, because the QI-groups with GID = 1, 3 and 4 have attribute age at most 35. However, if we are given an additional counterfeit table (Table 4.8(b)), we know that there is a counterfeit tuple in the QI-group with GID = 1 and another counterfeit tuple in the QI-group with GID = 4. Thus, we derive that the total number of individuals with age at most 35 is equal to 4 (= 6 - 2) tuples, which is accurate.

Algorithm

Now, we give the algorithm that generates an anonymzied table T_n^* from a table T_n under the privacy model of m-invariance. There are two cases.

- **Case 1:** T_n is the first release (i.e. $n = 1$).

Table 4.8: Published tables of Table 4.6(b) with counterfeits

(a) Published table T_2^* with counterfeits

Name	GID	Age	Zipcode	Disease
Alice	1	[25-26]	[54320-54321]	HIV
c_1	1	[25-26]	[54320-54321]	fever
Clement	2	[32-34]	[12350-12352]	cancer
David	2	[32-34]	[12350-12352]	flu
Emily	3	[47-49]	[35214-35214]	Heart Disease
Gary	3	[47-49]	[35214-35214]	flu
Helen	4	[22-26]	[54200-54211]	Cancer
c_2	4	[22-26]	[54200-54211]	flu

(b) Published counterfeit table

GID	Count
1	1
4	1

- **Case 2:** T_n is the second release or one of the later releases (i.e., $n \neq 1$).

Consider Case 1. We adopt one of the existing algorithms for l-diversity discussed in Section 2.3 where l is set to m, and modify data T_n to T_n^*. For example, consider 2-invariance. Table 4.6(a) is modified to Table 4.7(a) that satisfies l-diversity where $l = 2$.

Case 2 is more complicated. Consider that we published tables $T_1^*, T_2^*, ..., T_{n-1}^*$, and we want to modify T_n to T_n^* for data publishing. Under the m-invariance model, in order to generate T_n^*, we just need to read two tables: (1) T_n and (2) the table that was just previously published (i.e., T_{n-1}^*). For example, in our running example, if we want to generate T_2^*, we just need to read T_2 (Table 4.6(a)) and T_1^* (Table 4.7(a)).

There are some insertion operations and some deletion operations. For example, in our running example, the set of individuals (or tuples) in T_{n-1}^* is different from the set of individuals in T_n. The set S_1 of individuals in T_1^* is {Alice, Bob, Clement, David, Emily, Fred, Helen, Iris} and the set S_2 of individuals in T_2 is {Alice, Helen, Clement, David, Emily, Gary}.

Let $S_\cap$ be the set of individuals that appear in both S_n and S_{n-1}, and S_- be the set of individuals that appear in S_n but not in S_{n-1}. For instance, $S_\cap$ is equal to {Ailce, Clement, David, Emily, Helen} and S_- is equal to {Gary}.

There are the following three steps to generate T_n^*.

- **Step 1 (Division):** We partition $S_\cap$ into a number of *buckets*, each of which contains only individuals with the same signature. For example, in Figure 4.3(a), we generate three different buckets, namely B_1, B_2 and B_3, corresponding to different signatures. Then, we insert different individuals in S_{cap} into different buckets according to their signatures. For example, bucket B_1

is associated with signature {HIV, fever}. Since Alice's signature is {HIV, fever}, she is inserted into bucket B_1. The resulting partitioning on $S_\cap$ is shown in Figure 4.3(a).

- **Step 2 (Assignment):** A bucket is *balanced* if the total number of individuals in the bucket is exactly a multiple of the total number of sensitive values in the signature associated with the bucket. For example, in Figure 4.3(a), in bucket B_1, there is only 1 individual (namely, Alice), but there are two sensitive values (namely, HIV and fever). Thus, B_1 is not balanced. However, if there is an additional individual in B_1, then B_1 would be balanced. Similarly, bucket B_2 and bucket B_3 are not balanced. If B_2 (B_3) contains one additional individual, it is balanced.

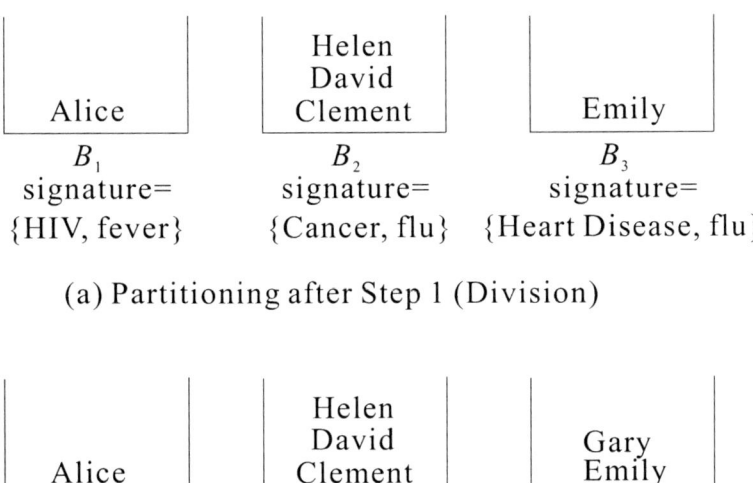

(a) Partitioning after Step 1 (Division)

(b) Partitioning after Step 2 for S.

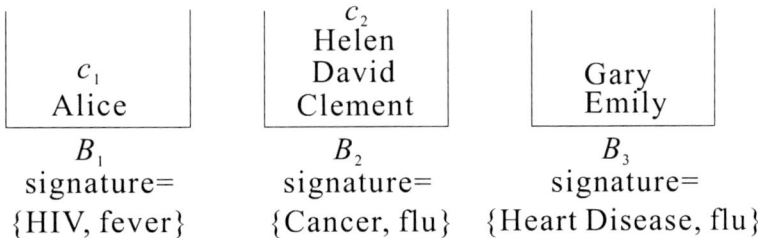

(c) Partitioning after Step 2 for counterfeit tuples

Figure 4.3: Illustration of generating a table under the m-invariance privacy model

In this step, we need to insert the individuals in S_- and some counterfeit tuples (or individuals) into different buckets such that each bucket is balanced.

Firstly, we insert the individual in S_- into buckets according to the sensitive values of individuals in S_-. Each individual in S_- is inserted into the bucket such that the sensitive value of the individual appears in the signature associated with the bucket. For example, since Gary is in S_- and his sensitive value is flu, we insert him into bucket B_3 because flu appears in the signature associated with B_3 (i.e., {Heart Disease, flu}). Figure 4.3(b) shows the partitioning after this step.

Secondly, we insert a number of counterfeit tuples into the buckets such that each bucket is balanced. Consider our running example. Since bucket B_1 and bucket B_2 are not balanced, we insert a counterfeit tuple c_1 into B_1 and another counterfeit tuple c_2 into B_2.

Figure 4.3(c) shows the partitioning after we insert counterfeit tuples.

- **Step 3 (Split):** After we obtain the partitioning from Step 2, we generate T_n^* as follows. For each bucket (which is balanced), we generate multiple QI-groups of size n where n is the total number of sensitive values in the signature associated with the bucket.

  Consider Figure 4.3(c). For bucket B_1, we generate a QI-group containing Alice and c_1. For bucket B_2, we generate two QI-groups. One group contains Clement and David, and the other group contains Helen and c_2. For bucket B_3, we generate a QI-group containing Emily and Gary.

  Finally, according to these QI-groups, we generate the anonymized table and the counterfeit table as shown in Table 4.8.

4.2 SENSITIVE VALUE-BASED CORRELATION

In the previous section, we assume that the QI attributes and the sensitive attribute of each individual do not change. However, in many real applications, both attributes change over time.

There are the following main challenges when the data can be updated with the QI attributes and the sensitive attributes.

- Firstly, both the QI value and sensitive value of an individual can change. For example, after a move, the postal code of an individual changes. That is, the external table such as a voter registration list can have multiple releases and changes from time to time. Also, a patient may recover from one disease but develop another disease.

- Secondly, some special sensitive values should remain unchanged. That is, some sensitive values that once linked to an individual can never be unlinked. For instance, in medical records, sensitive diseases such as HIV, diabetes and cancers are, to this date, not completely curable, and, therefore, they are expected to persist. We call these values *permanent sensitive values*.

Table 4.9: Voter Registration List(RL) and Microdata(T)

PID	Age	Zip.
p_1	23	16355
p_2	22	15500
p_3	21	12900
p_4	26	18310
p_5	25	25000
p_6	20	29000
p_7	24	33000
...	...	...
$p_{|RL|}$	31	31000

(a) $\mathcal{RL}_1$

PID	Age	Zip.
p_1	23	16355
p_2	22	15500
p_3	21	12900
p_4	26	18310
p_5	25	25000
p_6	20	29000
p_7	24	33000
...	...	...
$p_{|RL|}$	31	31000

(b) $\mathcal{RL}_2$

PID	Age	Zip.
p_1	23	16355
p_2	22	15500
p_3	21	12900
p_4	26	18310
p_5	25	*15000*
p_6	20	29000
p_7	24	33000
...	...	...
$p_{|RL|}$	31	31000

(c) $\mathcal{RL}_3$

PID	Disease
p_1	Flu
p_2	HIV
p_3	Fever
p_4	HIV
p_5	Flu
p_6	Fever

(d) T_1

PID	Disease
p_1	Flu
p_2	HIV
p_3	*Flu*
p_4	HIV
p_5	*Fever*
p_6	Fever

(e) T_2

PID	Disease
p_1	Flu
p_2	HIV
p_3	Flu
p_4	HIV
p_5	Fever
p_6	Fever

(f) T_3

Permanent sensitive values can be found in many domains of interest. Some examples are having a pilot's qualification and having a criminal record.

We use m-invariance as a representative to illustrate the inadequacy of traditional approaches for the above scenarios, by the following example. In Table 4.9, $\mathcal{RL}_1$, $\mathcal{RL}_2$ and $\mathcal{RL}_3$ are snapshots of a voter registration list at times t_1, t_2 and t_3, respectively. Note that Zipcode of p_5 is updated in $\mathcal{RL}_3$. The microdata table T_1, T_2 and T_3 are to be anonymized at times t_1, t_2 and t_3, respectively. The corresponding anonymized tables (T_1^*, T_2^* and T_3^*) are shown in Table 4.10. It is easy to see that T_1^*, T_2^* and T_3^* satisfy 3-invariance. This is because, in any release, for each individual, the set of 3 distinct sensitive values that the individual is linked to in the corresponding QI-group remains unchanged. Note that HIV is a *permanent* disease, but Flu and Fever are *transient* diseases. Furthermore, assume that from the registration lists, one can determine that $p_1, p_2, ..., p_6$ are the only individuals who satisfy the QI conditions for the groups with GID = 1 and GID = 2 in all the three tables of T_1^*, T_2^* and T_3^*. Then, surprisingly, the adversary can determine that p_4 has HIV with 100% probability. The reason is based on *possible world exclusion* from all published releases. First, we show that p_1 and p_6 cannot be linked to HIV. Suppose that p_1 suffers from HIV. In T_1^*, since p_1, p_2 and p_3 form a QI-group containing one HIV value, we deduce that neither p_2 nor p_3 are linked to HIV.

Table 4.10: Published Tables T^* satisfying 3-invariance

PID	GID	Age	Zip.	Disease
p_1	1	[21, 23]	[12k, 17k]	Flu
p_2	1	[21, 23]	[12k, 17k]	HIV
p_3	1	[21, 23]	[12k, 17k]	Fever
p_4	2	[20, 26]	[18k, 29k]	HIV
p_5	2	[20, 26]	[18k, 29k]	Flu
p_6	2	[20, 26]	[18k, 29k]	Fever

(a) First Publication T_1^*

PID	GID	Age	Zip.	Disease
p_2	1	[20, 22]	[12k, 29k]	HIV
p_3	1	[20, 22]	[12k, 29k]	Flu
p_6	1	[20, 22]	[12k, 29k]	Fever
p_1	2	[23, 26]	[16k, 25k]	Flu
p_4	2	[23, 26]	[16k, 25k]	HIV
p_5	2	[23, 26]	[16k, 25k]	Fever

(b) Second Publication T_2^*

PID	GID	Age	Zip.	Disease
p_2	1	[21, 25]	[12k, 16k]	HIV
p_3	1	[21, 25]	[12k, 16k]	Flu
p_5	1	[21, 25]	[12k, 16k]	Fever
p_1	2	[20, 26]	[16k, 29k]	Flu
p_4	2	[20, 26]	[16k, 29k]	HIV
p_6	2	[20, 26]	[16k, 29k]	Fever

(c) Third Publication T_3^*

Similarly, in T_2^*, since p_1, p_4 and p_5 form a QI-group containing one HIV value, p_4 and p_5 do not have HIV values. Also, from T_3^*, we deduce that p_4 and p_6 are not linked to HIV. Then, we conclude that p_2, p_3, p_4, p_5 and p_6 do not have HIV. However, we know that there are two HIV values in each of the releases T_1^*, T_2^* and T_3^*. This leads to a contradiction. Thus, p_1 cannot be linked to HIV. By similar induction, p_6 cannot be an HIV carrier. Finally, from the QI-group with GID = 2 in T_3^*, we figure out that p_4 must be an HIV carrier! No matter how large m is, this kind of possible world exclusion can appear after several rounds of publishing. Note that even if the registration list remains unchanged, the same problem can occur since the six individuals can be grouped in the same way as in T_1^*, T_2^* and T_3^* at 3 different times, according to the algorithm proposed by Xiao and Tao [2007] (described in Section 4.1.2).

- Thirdly, the anonymization mechanism for serial publishing should provide *individual-based protection* in which the probability that an *individual* is linked to a sensitive value is bounded by a privacy threshold. Yet, previous works (e.g., [Byun et al., 2006; Xiao and Tao, 2007]) focus on *record-based protection* in which the probability that a *record* is linked to a sensitive value is bounded by a privacy threshold. Since an individual can own *multiple* records in different releases, record-based protection is insufficient to protect *individual* privacy. Specifically, in m-invariance [Xiao and Tao, 2007], each record is associated with a lifespan of contiguous releases and a signature that is an *invariant* set of sensitive values linking to record r_j in the published table. If a record r_j for individual p_i appears at time t_j, disappears at time t_{j+1} (e.g. p_i may discontinue treatment or may switch to another hospital), and reappears at time t_{j+2}, the appearance at t_{j+2} is treated as a new record r_{j+2} in the anonymization process adopted by Xiao and Tao [2007]. There is no memory of the previous signature for r_j, and a new signature is created for r_{j+2}. Let us take a look at T_1^* in Table 4.10. From T_1^*, we can find that by 3-invariance, the signature of the records for p_1 and p_3 in T_1^* is {Flu, HIV, Fever}. If p_1 and p_3 recover from Flu and Fever at time t_2 (thus absent in T_2), and reappear due to other disease at time t_3 (in T_3), the reappearance of p_1 and p_3 in T_3 is treated as new records r_1', r_3' and by m-invariance, there is no constraint for their signature. Thus, at time t_3, if the signatures for r_1' and r_3' do not contain HIV, p_1 and p_3 will be excluded from HIV. Consequently, p_2 will be found to have HIV! It is not easy to extend m-invariance to individual-based protection. For example, binding invariant signatures to individuals is not feasible since an individual may develop new diseases that are not in the signature.

- Fourthly, the knowledge model for the adversary should be realistic. Existing works assume that it is trivial for the adversary to obtain the background knowledge of each individual's presence or absence in every snapshot of the microdata. However, gaining this kind of *participation knowledge* can be as hard as knowing the individual's sensitive values because one's participation in a microdata snapshot is also confidential. For example, Nergiz et al. [2007] deal with protecting the information about the presence of individuals in a data release. A more plausible scenario is that an adversary knows the participation knowledge of a few close friends.

As we mentioned, there are two kinds of values in the sensitive attribute, namely *permanent sensitive values* and *transient sensitive values*. In the following, we give a method that protects individual privacy for permanent sensitive values (Section 4.2.1) and another method that protects individual privacy for transient sensitive values (Section 4.2.2).

4.2.1 PROTECTION FOR PERMANENT SENSITIVE VALUES

Bu et al. [2008] propose an anonymization method called *HD-composition* that protects individual privacy for permanent sensitive values. The method involves two major roles, namely *holder* and

84 4. MULTIPLE-TIME DATA PUBLISHING

decoy. The objective is to bound the probability of linkage between any individual and any permanent sensitive value by a given threshold, e.g., $1/\ell$. Suppose an individual p_i has a permanent sensitive value s in the microdata. One major technique used for anonymizing static data is to form a QI-group mixing p_i and other individuals whose sensitive values are not s. By merely having the published QI-groups, the adversary cannot establish strong linkage between p_i and s. The anonymization also follows this basic principle, where the individual to be protected is named as a holder and some other individuals for protection are named as decoys.

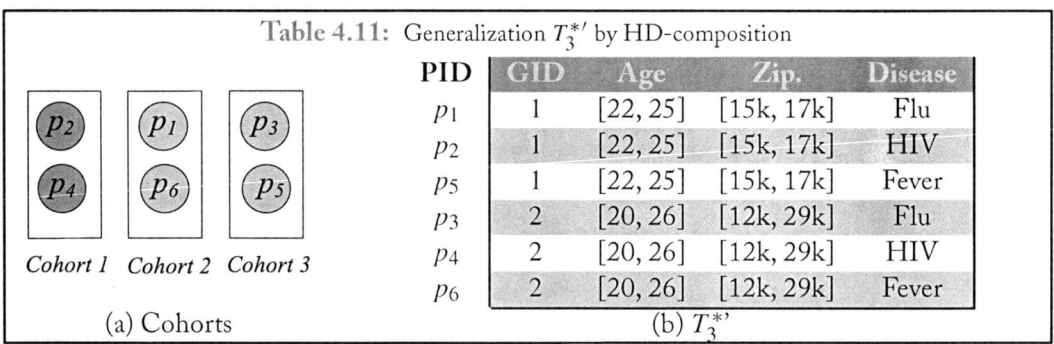

Table 4.11: Generalization $T_3^{*\prime}$ by HD-composition

(a) Cohorts

(b) $T_3^{*\prime}$

There are two major principles for privacy protection: *role-based partition* and *cohort-based partition*. By role-based partition, in every QI-group of the published data, for each holder of a permanent sensitive value s, we find $\ell - 1$ decoys that are not linked to s. Thus, each holder is masked by $\ell - 1$ decoys. In cohort-based partitioning, for each permanent sensitive value s, the proposed method constructs ℓ cohorts, one for holders and the other $\ell - 1$ for decoys. There is a restriction that decoys from the same cohort cannot be placed in the same partition – this is to imitate the properties of true holders.

The table generated according to these two major principles can protect individual privacy over multiple releases when there exists permanent sensitive values. That is, the probability of linkage between any individual and any permanent sensitive value is bounded by a given threshold $1/\ell$ where ℓ is a user input parameter.

Consider the example in Table 4.9 and Table 4.10 where $\ell = 3$. Since p_2 and p_4 are HIV-holders in T_1. In Table 4.11(a), they are both in cohort 1 where all HIV-holders are stored. In T_1^*, p_1 and p_3 form a QI-group (the QI-group with GID = 1) with an HIV-holder (i.e. p_2), and they are HIV-decoys. p_1 and p_3 are inserted into cohort 2 and 3, respectively. Similarly, p_6 and p_5 are decoys for p_4 in T_1^*, and are inserted into cohorts 2 and 3, respectively. With the constraints of cohort-based partition, we get anonymized table $T_3^{*\prime}$ (Table 4.11(b)) rather than the problematic T_3^* (Table 4.10(c)). This is because in T_3^*, decoys p_1 and p_6 are grouped with holder p_4, but p_1 and p_6 are from cohort 2, which violates the constraint of cohort-based partition.

4.2.2 PROTECTION FOR TRANSIENT SENSITIVE VALUES

Wong et al. [2010] consider how to protect individual privacy for transient sensitive values. Let us consider a transient sensitive disease chlamydia, which is easily curable. Suppose that there exist 3 records of an individual p in 3 different medical data releases. It is obvious that typically p would not want anyone to deduce with high probability from these released data that s/he has ever contracted chlamydia in the past. Here, the past practically corresponds to *one or more* of the three data releases. Therefore, if from these data releases, an adversary can deduce, with high probability that p has contracted chlamydia in one or more of the three releases, privacy would have been breached. To protect privacy, we would like the probability of any individual being linked to a sensitive value in one or more data releases to be bounded from above by $1/\ell$. Let us call this privacy guarantee the *global guarantee* and the value $1/\ell$ the *privacy threshold*.

Though the global guarantee requirement seems to be quite obvious, it does not appear to have been considered so far. The closest guarantee provided by existing works is the following: for *each* of the data releases, p can be linked to chlamydia with a probability of no more than $1/\ell$. Let us call this guarantee the *localized guarantee*. Would this guarantee be equivalent to the global guarantee? In order to answer this question, let us look at an example.

Table 4.12: A motivating example

PID	Sex	Zipcode	Disease
p_1	M	65001	flu
p_2	M	65002	chlamydia
p_3	F	65014	flu
p_4	F	65015	fever

(a) T_1

PID	Sex	Zipcode	Disease
p_1	M	65001	chlamydia
p_2	M	65002	flu
p_3	F	65014	fever
p_5	F	65010	flu

(b) T_2

Table 4.13: Anonymization for T_1 and T_2

Sex	Zipcode	Disease
M	6500*	flu
M	6500*	chlamydia
F	6501*	flu
F	6501*	fever

(a) T_1^*

Sex	Zipcode	Disease
M	6500*	chlamydia
M	6500*	flu
F	6501*	fever
F	6501*	flu

(b) T_2^*

Consider two raw medical tables (or microdata) T_1 and T_2 as shown in Table 4.12 at time points t_1 and t_2, respectively. Suppose that they contain records for five individuals p_1, p_2, p_3, p_4, p_5. In this example, Sex and Zipcode are the quasi-identifier attributes, while Disease is the sensitive attribute. Attribute PID is used for illustration purpose and does not appear in the published table. We assume that each individual owns at most one tuple in each table at each time point. Furthermore, we assume no additional background knowledge about the linkage of individuals to diseases, and the sensitive values linked to individuals can be freely updated from one release to the next. For the sake of illustration, assume that all sensitive values are transient.

Table 4.14: Possible worlds for G_1 and G_2

Sex	Zipcode	Disease
M	65001	flu
M	65002	chlamydia

T_1

Sex	Zipcode	Disease
M	65001	flu
M	65002	chlamydia

T_2

(a) Possible world 1 w_1

Sex	Zipcode	Disease
M	65001	flu
M	65002	chlamydia

T_1

Sex	Zipcode	Disease
M	65001	chlamydia
M	65002	flu

T_2

(b) Possible world 2 w_2

Sex	Zipcode	Disease
M	65001	chlamydia
M	65002	flu

T_1

Sex	Zipcode	Disease
M	65001	flu
M	65002	chlamydia

T_2

(c) Possible world 3 w_3

Sex	Zipcode	Disease
M	65001	chlamydia
M	65002	flu

T_1

Sex	Zipcode	Disease
M	65001	chlamydia
M	65002	flu

T_2

(d) Possible world 4 w_4

Suppose the probability that an individual is linked to a transient sensitive value is at most $1/\ell$. We set $\ell = 2$. In a typical data anonymization [LeFevre et al., 2005, 2006; Sweeney, 2002b; Xiao and Tao, 2006a], in order to protect individual privacy, the QI attributes of the raw table are *generalized* or *bucketized* in order to form some *anonymized groups* to hide the linkage between an individual and a sensitive value. For example, T_1^* in Table 4.13(a) is a *generalized* table of T_1 in Table 4.12. We generalize the zipcode of the first two tuples to 6500* so that they have the same QI values in T_1^*. We say that these two tuples form an *anonymized group*. It is easy to see that in each published table (T_1^* or T_2^*), the probability of linking any individual to chlamydia or flu is at most 1/2, which satisfies the localized guarantee. The question is whether this satisfies the global privacy guarantee with a threshold of 1/2.

For the sake of illustration, let us focus on the anonymized groups G_1 and G_2 containing the first two tuples in T_1^* and T_2^* (Table 4.13), respectively. The probability in serial publishing can be derived by the *possible world analysis*. There are four possible *worlds* for G_1 and G_2 in these two published tables, as shown in Table 4.14. Here each *possible world* is one possible way to assign the diseases to the individuals in such a way that is consistent with the published tables. Therefore, each possible world is a possible assignment of the sensitive values to the individuals at all the publication time points for groups G_1 and G_2. Note that an individual can be assigned to different values at different data releases, and the assignment in one data release is independent of the assignment in another release.

Consider individual p_2. Among the four possible worlds, three link p_2 to chlamydia, namely w_1, w_2 and w_3. In w_1 and w_2, the linkage occurs at T_1, and in w_3, the linkage occurs at T_2. Thus, the probability that p_2 is linked to chlamydia in at least one of the tables is equal to 3/4, which is greater than 1/2, the intended privacy threshold. From this example, we can see that localized guarantee does not imply global guarantee.

Table 4.15: Anonymization for global guarantee

Sex	Zipcode	Disease	Sex	Zipcode	Disease
*	650**	flu	*	650**	chlamydia
*	650**	chlamydia	*	650**	flu
*	650**	flu	*	650**	fever
*	650**	fever	*	650**	flu
(a) T_1*			(b) T_2*		

Interestingly, Wong et al. [2010] show that global guarantee implies localized guarantee. They also show that in order to ensure the global guarantee, the sizes of the anonymized groups typically need to be larger than that needed for localized guarantee. In the above example, we can use size 4 anonymized groups as shown in Table 4.15. There will be 4! × 4! possible worlds. It is easy to see that 3/4 of the possible worlds do not assign chlamydia to p_2 in the first release, 3/4 of them do not assign chlamydia to p_2 in the second release, and $3/4 \times 3/4 = 9/16$ of the possible worlds do not assign chlamydia to p_2 in both releases. The remaining possible worlds assign chlamydia to p_2 in at least one of the two releases. Hence, the privacy breach probability is $1 - 9/16 = 7/16 < 1/2$. The exact condition for privacy guarantee is not simply a size requirement, but a bound on a size ratio between that of the group and the sensitive value occurrences. They also propose some methods that generate anonymized tables satisfying the global privacy guarantee. Details can be found in [Wong et al., 2010].

4.3 CONCLUSION

In this chapter, we learnt that data can be published *more than once*. We call this process *multiple-time data publishing* in contrast to one-time data publishing studied in Chapter 3. The techniques originally designed for one-time data publishing is insufficient when we consider multiple-time data publishing. This is because multiple-time data publishing introduces an additional aspect, namely time, which does not exist in one-time data publishing. Specifically, there are some correlations among different published data released at different times. For example, if an individual is linked to HIV in a previous release, s/he is also linked to HIV in all later releases. By considering the correlation, the data publisher has to modify the current data to be released so that individual privacy can be protected.

In this chapter, we studied two types of correlations for multiple-time data publishing. The first type is individual-based correlation (Section 4.1), and the second type is sensitive value-based correlation (Section 4.2). In individual-based correlation, a single individual can appear in multiple releases. In sensitive value-based correlation, the sensitive value of an individual can keep unchanged over multiple releases. According to these correlations, the adversary may breach individual privacy. Thus, the data publisher generate the modified data by avoiding privacy breaches.

CHAPTER 5

Graph Data

In previous chapters, we studied the privacy issues of publishing relational tables. In this chapter, we consider the same issues in publishing graph data. Graph data has emerged as a significant topic in recent studies of database systems. One important source of graph data is social network data. Facebook and MySpace are two typical examples of social networks. Publishing social network data can help investigations on the nature of the networks, but at the same time, it may introduce the risk of individual privacy breaches. A vivid example of this dilemma is the release of the Enron corpus. When Enron Corporation filed for bankruptcy in 2001, due to the fraud investigation involved, the corpus was made public by the Federal Energy Regulatory Commission. However, this dataset contained email correspondences of the employees and privacy became an issue. Publishing such data can be very valuable for researchers interested in how emails are used in an organization and better understanding of organizational structure. Since the privacy problem was identified, there have been a number of studies on how to anonymize the social network data so that privacy can be protected. In this chapter, we shall give an overview of some of the related issues and related works.

5.1 DATA MODEL

Different types of data graphs have been considered in the literature of privacy preserving publication of graph data. The simplest consideration is an unlabeled graph $G = (V, E)$, where V is a set of nodes and E is a set of edges. Typically, each node corresponds to an individual or a party whose privacy should be protected. An edge represents a relationship between two individuals or two parties.

In other models, vertices are assigned labels [Zhou and Pei, 2008]. For example the labels may correspond to the occupations of individuals.

The model for the Enron corpus would be a graph where each node corresponds to an individual, and there is a set of sensitive information or content attached with each node, such as the emails sent or received by the individual. A main difference between labels and contents is that identical labels can be common in the graph, while contents are more complex and two different nodes are expected to have different sets of contents. Also, one may want to publish the labels intact while sensitive information such as names or identities should be eliminated from the contents before publication.

Other than sensitive information at each node, edges in the data graph can also be sensitive. For example, we may consider that the dataset is a multi-graph $G = (V, E_1, E_2, ..., E_s)$ where V is a set of nodes, and E_i is a set of edges that belongs to a certain class. E_s corresponds to the sensitive relationships. Edges can also be labeled with information about the relationships. Figure

5.1 shows a small data graph where each node corresponds to an individual and the links represent the relationships among individuals.

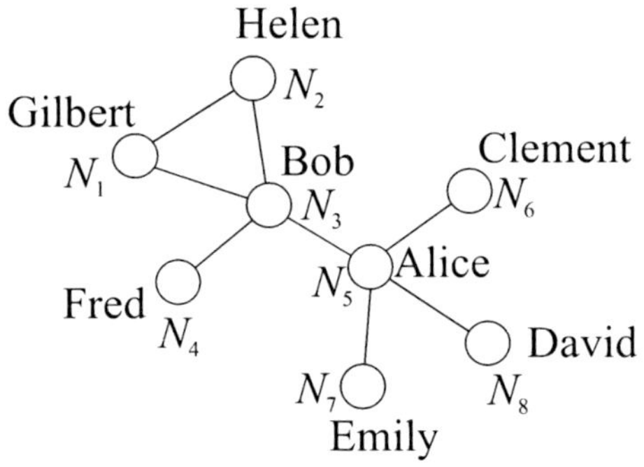

Figure 5.1: An example showing a network

5.2 ADVERSARY ATTACKS

The problem definition can differ in terms of the kinds of adversary attacks that are considered. In most works, privacy is considered to be breached when the adversary can isolate a node or vertex in the graph that is mapped to an individual. This is also called *node re-identification*.

Another possible target of attack is the linkages among individuals, since relationships with others may be considered sensitive information for some individuals.

An adversary can make use of the knowledge s/he possesses concerning the data graph to launch an attack. Hence part of the problem definition includes the assumption of adversary knowledge.

5.2.1 ASSUMPTION OF ADVERSARY KNOWLEDGE

A powerful knowledge of the adversary is *structural knowledge* related to a target individual. One type of structural knowledge specific to graph databases is related to the node degrees and a second type is subgraph knowledge.

1. Node Degrees:

    Hay et al. [2007] formulate the knowledge of an adversary as answers to certain classes of queries. Different query classes $\mathcal{H}_i(x)$ are defined for a node x. The meanings of such queries are as follows.

    - $\mathcal{H}_0(x)$ returns the label of a node x (in case the nodes are not labeled, $\mathcal{H}_0(x) = \epsilon$). Note that when $\mathcal{H}_0(x) = \epsilon$, it means that the adversary has no knowledge about the graph.
    - $\mathcal{H}_1(x)$ returns the degree of node x. With this query class, the adversary is assumed to know the degree of the target individual.

- $\mathcal{H}_2(x)$ returns the multiset of the degrees of each neighbor of x,
- ... and so on.

With the data graph given in Figure 5.1, $\mathcal{H}_0(N_5) = $ "$Alice$". If we remove the name labels, $\mathcal{H}_0(N_5) = \epsilon$. $\mathcal{H}_1(N_5) = 4$, $\mathcal{H}_2(N_5) = \{1, 1, 1, 4\}$.

2. Subgraph Queries

   A stronger and more realistic class of query is subgraph queries, which asserts the existence of a subgraph around the target node [Hay et al., 2008]. The descriptive power of the query is modeled by the number of edges in the subgraph.

   An example is shown in Figure 5.1. In the social network as shown in this figure, each node corresponds to an individual and each edge connecting two nodes corresponds to a close relationship between two corresponding individuals. Each node is associated with a node label, a node ID and multiple node attributes. In this example, the node label corresponds to the name of the corresponding individual. The node ID (denoted by N_i) is used for illustration purposes. Node attributes of a node N are some descriptors of N. Some examples of node attributes are Occupation, Gender and Age. To simplify our presentation, we do not consider node attributes in the following discussion.

   A straightforward way to publish the data is to remove the node labels as shown in Figure 5.2. However, such an anonymized network can still breach individual privacy. Consider that the adversary knows that Alice has four close friends. S/he also knows that none of these friends knows Alice's other friends. The adversary can figure out that node N_5 in Figure 5.2 corresponds to Alice. Similarly, the adversary may also deduce that node N_3 corresponds to Bob. Then, the adversary deduces that there is an edge between Alice and Bob. This breaches Alice's and Bob's privacy.

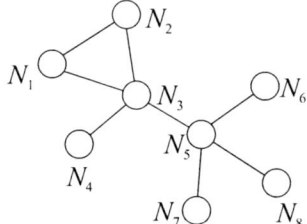

Figure 5.2: An example showing a network without users' identities

In the above example, the subgraph query representing the knowledge of the adversary can be formulated as the subgraph formed by the immediate neighbors of a node. This special kind of subgraph query is considered by Zhou and Pei [2008]. They assume that the adversary has the knowledge about the 1-neighborhood graph of each node (or individual). Given a node

N, the 1-neighborhood graph of node N is defined to be a subnetwork (or subgraph) of the original network where

- the set X of nodes in this subnetwork is the set of node N plus the *immediate neighbors* of N, i.e., nodes which have an edge connecting to node N,
- the set of edges in this subnetwork is the set of all edges in the original network containing both end-points in X.

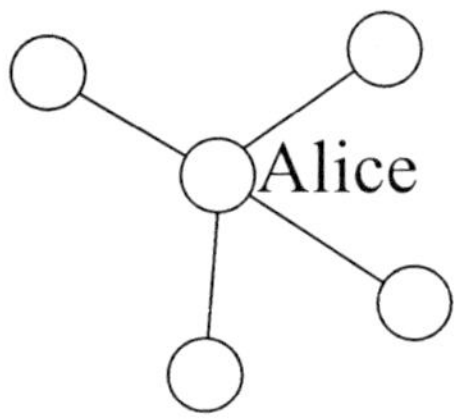

Figure 5.3: An example showing the 1-neighborhood graph of Alice

For example, Figure 5.3 shows the 1-neighborhood graph of Alice. However, 1-neighborhood places a lot of restriction on the adversary knowledge, and this assumption may not hold in general. A more powerful model will be not to limit the subgraph knowledge of the adversary. In this case, the subgraph can be of any size. Some recent related works are proposed in [Hay et al., 2008] and [Zou et al., 2009] and they will be described in Section 5.4.

5.2.2 ACTIVE ATTACKS

The terms passive versus active attacks are used to differentiate between cases where the adversary only observes the data and does not temper with the data versus the cases where the adversary may change the data in order to attack.

Consider active attacks aiming for identify disclosure or node re-identification [Backstrom et al., 2007]. In one possible scenario, the adversary would find or create k nodes in the network, namely $\{x_1, ..., x_k\}$, and then create each edge (x_i, x_j) independently with probability 1/2. This produces a random graph H, which will be merged with the given social network forming a graph G. The first requirement of H is that there is no other subgraph S in G that is isomorphic to H (meaning that H can result from S by relabeling the nodes). In this way, H can be uniquely identified in G. The second requirement of H is that there is no automorphism, which is an isomorphism from H to itself. With H, the adversary can link a target node w to a subset of nodes N in H so that once H is located, w is also located. From results in random graph theory [Bollobas, 2001], the chance of achieving the two requirements above is very high by using the random graph generation method.

Note that if the attack is based on a subgraph in the network, it can be defended in the same manner as for passive attack based on subgraph knowledge.

5.3 UTILITY OF THE PUBLISHED DATA

From the above discussion, we can see that publishing the raw data is problematic even if the individual labels or id's are removed. For privacy preservation, typically the raw data would be distorted in some way to prevent the leakage of sensitive information. The distortion may cause damage to the utility of the data. Hence it is necessary to minimize the distortion in such a way as to preserve a good level of utility. Before we introduce mechanisms for data anonymization, we consider the utility of the published data.

There can be different kinds of utility for social network data. Kumar, Novak, Tomkins studied the structure and evolution of the network [Kumar et al., 2006]. A considerable amount of research has been conducted on social network analysis [Backstrom et al., 2006; Kossinets et al., 2008; Leskovec et al., 2008]. There are also other query processing on or knowledge discovery from social networks [Cheng et al., 2009; Faloutsos et al., 2004; Tong et al., 2007]. [McCallum et al., 2005] considers topic and role discovery in social networks. Though there are different applications, all the above usages of graph data depend on certain graph properties. Therefore, in most previous works, the experimental measurement of the quality or utility of the data after anonymization is based on the measurement of the distortion of certain major graph properties. The following are some common measurements that have been used.

- Properties of interest in network data: [Hay et al., 2008] uses the following properties for the utility measure.

    - Degree: distribution of degrees of all nodes in the graph
    - Path length: distribution of the lengths of the shortest paths between 500 randomly sampled pairs of nodes in the network
    - Transitivity: distributions of the size of the connected component that a node belongs to.
    - Network resilience: number of nodes in the largest connected component of the graph as nodes are removed
    - Infectiousness: proportion of nodes infected by a hypothetical disease, first randomly pick a node for the first infection, and then spread the disease with a specific rate.

- Aggregate query answering: Zhou and Pei [2008] consider queries of the following form: for two labels l_1, l_2, what is the average distance from a node with label l_1 to its nearest node with label l_2?

By comparing the measurements of the above properties before and after data anonymization, it can be shown how far the data utilities have been reserved.

Though the above measurements can be adopted in the empirical studies, they do not lend themselves to easy or simple formulation to be incorporated in the problem definition or anonymization techniques. Hence simpler utility models are defined based on the kind of distortion that may be introduced by the anonymization process. For example, [Liu and Terzi, 2008] alters the vertex degrees, and the utility model is based on the changes made in terms of the vertex degrees. In [Zou et al., 2009], the anonymization involves edge additions and vertex additions, and the anonymization cost for the utility model is defined as the edit distance between the original graph and the anonymized graph:

Definition 5.1 Given a graph G and its anonymized version $G*$, the anonymization cost in $G*$ is defined as
$$Cost(G, G*) = (E(G) \cup E(G*)) = (E(G) \cap E(G*))$$
where $E(G)$ is the set of edges in G.

It is with such cost measurements that the NP-completeness or NP-hardness of the problems in privacy preserving data publishing are proven. While a simpler utility model such as the vertex distortion or edit distance is used in the problem definition for minimization of distortion, in the empirical studies, the degree of preservation of data utilities is typically measured by the distortion introduced to the graph properties such as those listed in the above.

5.4 k-ANONYMITY

It is possible to prevent node re-identification by adopting the concept of k-anonymity. Specifically, with this knowledge, the data owner wants to modify the original network data G and generate another network data G^* to be published such that, in G^*, for each node A, there exists at least $k - 1$ other nodes that are not distinguishable from A based on adversary's knowledge.

There are different ways to modify the network data. Edge removals and edge insertions are common ways for this purpose. Another way is to form supernodes where each supernode is made up of at least k original nodes, which are then non-distinguishable.

5.4.1 VERTEX DEGREE

Liu and Terzi [2008] propose k-anonymity against vertex degrees; that is, every vertex v has at least $(k - 1)$ other vertices that have the same neighborhood subgraphs or vertex degrees as v. The adversary attack is based on the degree of the target node; that is, the knowledge of the adversary is given by $\mathcal{H}_1$. Given a graph $G = (V, E)$, if we sort the degrees of each node, then the sorted list is a degree sequence for G. Anonymization is based on the construction of a graph that follows a given degree sequence.

To preserve the utility of the published data, we measure the distance of a degree sequence $\hat{d}$ from the original sequence d by the following:
$$L_1(\hat{d} - d) = \sum_i |\hat{d}(i) - d(i)|$$
where $d(i)$ refers to the i-th value in the list d.

5.4. k-ANONYMITY

A dynamic programming algorithm is proposed that minimizes the above distance while constructing a degree sequence that is k-anonymous. Then a graph is generated that follows this sequence with an algorithm from Erdos and Gallai [1960].

As an example, consider the graph in Figure 5.2, the degrees for the nodes $N_1, N_2, ..., N_8$ are 3, 3, 4, 1, 4, 1, 1, 1, respectively. If we sort this list of degrees, we get the degree sequence $S = \{$ 4, 4, 3, 3, 1, 1, 1, 1 $\}$. This is 2-anonymous because for any given degree that exists in the sequence, the frequency of the degree is at least 2. Hence if an adversary knows the vertex degree of a target, there are at least 2 different vertices that can match the target. If 3-anonymity is needed, the sequence need to be anonymized, and a sequence that has minimal distance from S while satisfying 3-anonymity will be $\{3, 3, 3, 3, 1, 1, 1, 1\}$, with a distance of 2 from S.

5.4.2 1-NEIGHBORHOOD

Zhou and Pei [2008] consider the 1-neighborhood of a node as possible adversary knowledge and attempt to prevent node re-identification. In their data model, the vertices in the graph are labeled. The task is to modify the original network data G and generate another network data G^* to be published such that, in G^*, for each node A, there exists at least $k-1$ other nodes which are not distinguishable from A based on the 1-neighborhoods. Here graph changes by edge addition, and change of labels are allowed in the algorithm.

In our earlier example in Figure 5.2, after we add an edge between Emily and David, the k-anonymity requirement can be satisfied.

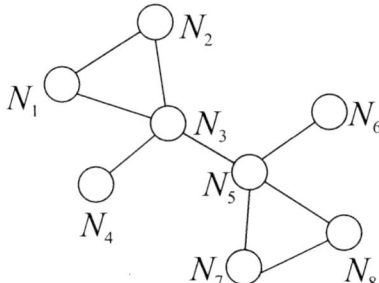

Figure 5.4: A graph which satisfies 2-anonymity

There are two main steps in the anonymization algorithm:

1. Find the 1-neighborhood graph of each node in the network.

2. Group the nodes by similarity and modify the neighborhoods of the nodes with the goal of having at least k nodes with the same 1-neighborhood in the anonymized graph.

The second step faces the difficult problem of graph isomorphism, which is one of a very small number of known problems in NP with no known polynomial algorithm and also no known proof of NP-completeness. Hence only exponential time algorithms are available. To solve this problem, Zhou and Pei [2008] adopts a minimum depth-first-search coding (DFS) method by Yan and Han [2002].

An outline of the method is the following. For a node N, let $G(N)$ be the 1-neighborhood graph. The induced graph for the set of immediate neighbors of N is partitioned into maximal connected subgraphs in $G(N)$. Such subgraphs are called the neighborhood components of N. For each such neighborhood component, the DFS method is used to encode the graph in a unique way. The DFS encodings of all the neighborhood components of N are combined into a canonical label for N. By comparing the canonical labels of two nodes, N_1 and N_2, we can determine if the 1-neighborhood for the two nodes are isomorphic.

5.4.3 VERTEX PARTITIONING

Hay et al. [2008] propose to protect privacy against subgraph knowledge by partitioning vertices and grouping each partition into a single vertex. By making sure that each partition contains at least k vertices, k-anonymity can be achieved with respect to attacks by any neighborhood subgraph.

The algorithm forms clusters of nodes of size at least k. The aim is to make the nodes inside each cluster indistinguishable, so that node re-identification is not possible. This method can resist attacks based on different kinds of adversary knowledge, including $\mathcal{H}_i$ and neighboring subgraph of any size.

Each cluster of nodes is called a *supernode*, and an edge (X, Y) between two supernodes X, Y is called a *superedge*. The superedges (X, Y) include self-loops and are labeled with a non-negative weight $d(X, Y)$, which indicates the density of edges within and across supernodes.

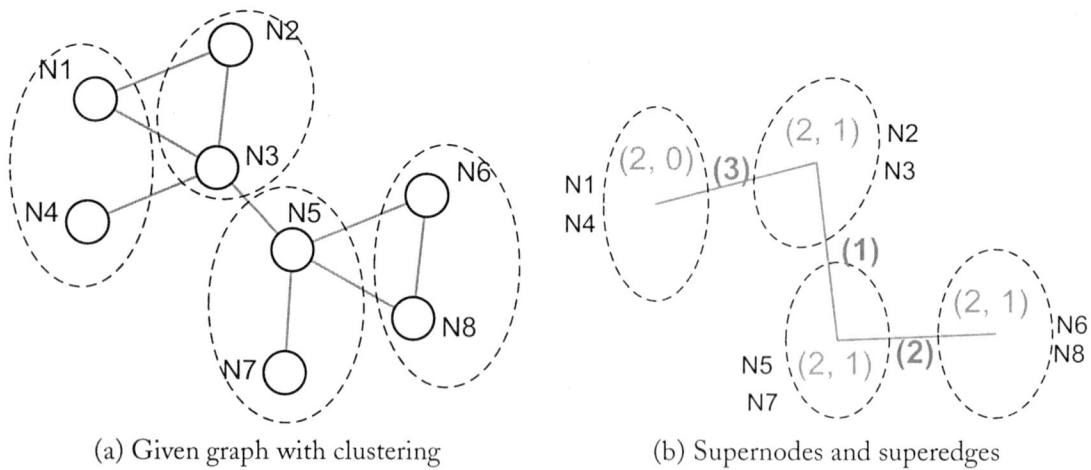

(a) Given graph with clustering   (b) Supernodes and superedges

Figure 5.5: An example showing graph partitioning for 2-anonymity

Figure 5.5(a) shows an example of graph partitioning for 2-anonymity. In the figure, each cluster is highlighted by a dotted line circle. There are two vertices in each cluster, or supernode. The supernodes are shown in Figure 5.5(b) with labels of the form (x, y) where x is the number of original vertices in the supernode, and y is the number of edges linking these vertices in the original graph. Each superedge links two clusters and is labeled with the number of edges across the two clusters in the original graph. For example, there are 3 edges in the original graph linking vertices in the supernode for (N1,N4) and the supernode for (N2,N3).

In order to preserve the utility, we examine the likelihood of recovering the original graph given the anonymized graph. If a graph $G = (V, E)$ with supernodes and superedges is published instead of the original graph, there will be multiple possibilities of the original graph; we consider each such possibility a possible world. Let $W(G)$ be the set of all possible worlds. If we assume equal probability for all possible worlds, then the probability of the original graph, which is one of the possible worlds, is given by $1/|W(G)|$.

We can formulate $|W(G)|$ by

$$|W(G)| = \prod_{X \in V} \binom{\frac{1}{2}|X|(|X|-1)}{d(X,X)} \prod_{X,Y \in V} \binom{|X||Y|}{d(X,Y)}$$

The value of $1/W(G)$ should be maximized so that the published graph is similar to the original graph since a higher value of $1/W(G)$ means that it is more likely to recover the original graph from the published graph. The algorithm follows a simulated annealing method to search for a good solution. Starting with a graph with a single supernode which contains all the original nodes, the algorithm repeatedly tries to find alternatives by splitting a supernode, merging two supernodes, or moving an original node from one supernode to another. If the above probability of $1/|W(G)|$ increases with the newly formed alternative, the alternative is accepted for further exploration. The termination criterion is that fewer than 10% of the alternatives are accepted.

Campan and Truta [2008] also protect privacy by forming clusters of vertices and collapsing each cluster into a single vertex. Here the problem setting is different from [Hay et al., 2008]. The network model includes information for each node in terms of attributes belonging to three categories:

- $I_1, ..., I_m$ are identifier attributes such as name and other identifiers

- $Q_1, ..., Q_q$ are quasi-identifier attributes such as zip_code and age that may be known by an adversary

- $S_1, ..., S_r$ are confidential or sensitive attributes such as medical diagnosis and income that are assumed to be unknown to an adversary.

As in [Hay et al., 2008], the published graph is made up of supernodes and superedges. However, since quasi-identifiers are assumed, they are generalized in each supernode to form a generalized tuple. Therefore, the information loss involved in the generalization will also be considered in forming the clustered graph.

5.4.4 k-AUTOMORPHISM

Zou et al. [2009] point out that the vertex clustering method in Hay et al. [2008] introduces ambiguity in the published data which may not be desirable for some applications. In order to achieve k-anonymization for any subgraph knowledge possessed by the adversary, they anonymize the data graph by edge and node additions so that the resulting graph is k-automorphic, which is defined as follows.

Definition 5.2 An automorphism of a graph $G = (V, E)$ is an automorphic function f of the vertex set V, such that for any edge $e = (u, v)$, $f(e) = (f(u), f(v))$ is also an edge in G.

Definition 5.3 Given a graph G, if there exist $k - 1$ automorphic functions $F_a (a = 1, ..., k - 1)$ in G and for each vertex v in G, $F_{a_1}(v) \neq F_{a_2}(v) 1 \leq a_1 \neq a_2 \leq k - 1$, then G is called a k-automorphic graph.

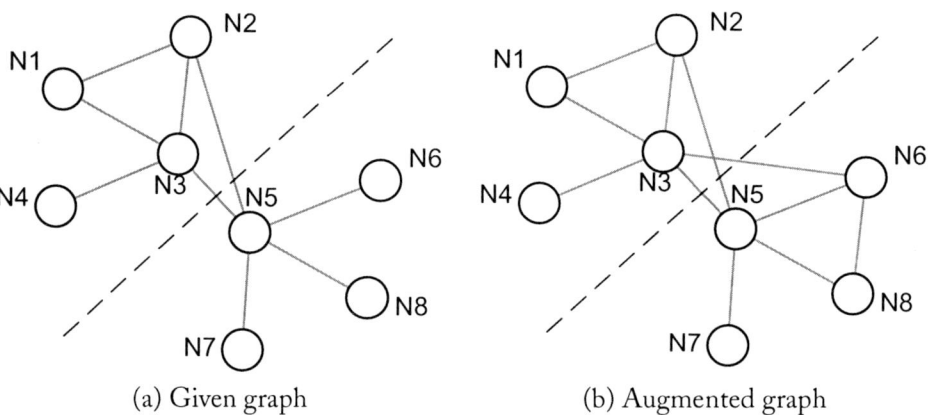

(a) Given graph (b) Augmented graph

Figure 5.6: Turning a given graph into a 2-automorphic graph

Figure 5.6(b) shows an example of a 2-automorphic graph. Here an automorphic function F can be defined as follows: $F(N1) = N8, F(N2) = N6, F(N3) = N5, F(N4) = N7, F(N5) = N3, F(N6) = N2, F(N7) = N4, F(N8) = N1$.

With k-automorphism, k-anonymity can be achieved with respect to subgraph attack based on any subgraph of the given graph. Suppose G is a graph with all user IDs removed. The K-Match (KM) algorithm in [Zou et al., 2009] partitions the graph G to form k blocks $B_1, ..., B_k$. with equal number of vertices. The vertices are mapped so that a mapping table called Alignment Vertex Table (AVT) can be formed. There are k columns $C_1, ..., C_k$ in AVT, each column contains all vertices from one block B_i. If r_k is a row in AVT, then the i-th element in r_k, $v(k, i)$ is a unique vertex in B_i, $v(k, i)$ is said to be mapped to $v(k, j)$ (which appears in the same row) in block B_j for $i \neq j$. All vertices in the blocks are mapped exactly once in AVT.

With the vertex mapping in AVT, the aim is to ensure that we can form $k-1$ automorphic functions F_{a_i} where given a vertex $v(i, k)$ in AVT,

$F_{a_1}(v(k, i)) = v(k, (i \mod k) + 1)$,
$F_{a_2}(v(k, i)) = v(k, (i + 1 \mod k) + 1)$, ...,
$F_{a_{k-1}}(v(k, i)) = (v(i, (i + k - 2 \mod k) + 1)$.

With $B_1, ..., B_k$, two conditions should be met in the anonymized graph G^*:

1. Edges within the blocks are matched; this is also called block alignment: for any edge between two vertices $v(j, i)$ and $v(k, i)$ in a block B_i, we must make sure that the corresponding edge $v(j, l)$ and $v(k, l)$ must exist for all other blocks B_l.

2. Edges crossing blocks are matched, this is ensured by edge copying: if any edge between vertices $v(i, k), v(j, k)$ exists between blocks B_i and B_j, then, the edges for the mapped vertices $v(r, k), v(s, k)$ must exist between any blocks B_r and B_s.

For example, if the given graph G is as shown in Figure 5.6(a), we can form 2 blocks $B_1 = \{N1, N2, N3, N4\}$ and $B_2 = \{N5, N6, N7, N8\}$. The corresponding AVT is shown in Figure 5.7.

N1	N8
N2	N6
N3	N5
N4	N7

Figure 5.7: Alignment Vertex Table (AVT)

However, the blocks do not match, since there is an edge between $N1$ and $N2$ in block B_1 but no edge between $N8$ and $N6$ in block B_2. The anonymization process will add the edge $(N8, N6)$ to the graph for block alignment. The resulting graph is still not 2-automorphic. There is a crossing edge linking $N2$ and $N5$, but there is no corresponding edge linking $N6$ and $N3$ exists. Therefore with the edge copying step, an edge $(N3, N6)$ is added, forming the graph in Figure 5.6(b). This graph is 2-automorphic as discussed above, and it can be published.

To facilitate graph partitioning, the algorithm first finds all sub-graphs with at least k matches in the graph G, where any two matches are edge-disjoint. Such subgraphs are called frequent subgraphs ([Kuramochi and Karypis, 2005]). The frequent subgraph with the most edges is chosen for partitioning, and its matches are extracted from G to form initial blocks for constructing G^*. Since these matches may not be vertex disjoint, dummy nodes are introduced to make them vertex disjoint. The blocks are then made to satisfy the above two conditions by edge additions.

After these blocks are formed, they are extended by including neighboring vertices recursively until each vertex in the graph belongs to some block. If a vertex is a neighbor of two or more blocks, it is randomly inserted into one of the blocks. Dummy vertices may need to be introduced to ensure graph isomorphism. Edges may be added to satisfy condition (1) in the above for the enlarged blocks.

After each expansion, edges may need to be introduced to satisfy both conditions given above. The expansion is repeated until all vertices in G are processed.

5.5 MULTIPLE RELEASES OF DATA GRAPHS

Most of the existing works consider only a one-time publication of data graphs. As noted earlier, data graphs are not static, but undergo changes over time. It will be useful to publish the data graph as it evolves; in which case, it is of interest to protect the privacy of users when the adversary has assess to all the releases of the graph.

This problem has been considered in Zou et al. [2009]. It is argued that multiple releases should be published in such a way that a node for the same individual can be traced over the history of releases since the data evolution can be interesting. Hence the vertices are labeled with unique IDs, and vertices corresponding to the same individual will preserve this ID over multiple releases. This, however, makes it possible for the adversary to launch privacy attacks.

A mechanism based on generalized IDs is proposed to tackle this problem. Assume that the KM algorithm for single-release privacy preservation is used to generate a k-automorphic graph. With multiple releases, the automorphism functions may undergo changes. Let G_t^* be the graph published at the t-th release. Let F_a^t be $k-1$ automorphic functions in the t-th release G_t^*. In order to maintain privacy, it is proposed that when $F_a^1(v) \neq F_a^t(v)$ for a vertex v, then a generalized vertex ID is created for $F_a^t(v)$, and vertex ID $F_a^1(v)$ is inserted into this generalized ID. Generalized vertex ID's are thus created whenever the above condition arises for all vertices. Next, each original vertex ID is replaced by the corresponding generalized ID, if any. It is shown that with this method of generalized vertex ID's, it is not possible to re-identify any vertex with a probability higher than $1/k$.

For vertex addition and deletion, if the adversary knows that a target user exists in one release but not in another release, it is possible to identify the target. To handle this case, the following steps are proposed [Zou et al., 2009]. Let G_t' be the k-automorphic graph generated from the t-th data graph based on the single release algorithm.

1. Vertex Deletion

   Suppose that due to vertex deletion, there is a vertex ID v which is in G_1^* but not in G_t'. In this case, an arbitrary vertex ID u that exists in both G_1' and G_t' is chosen, and v is inserted into the generalized vertex ID for u in G_t'.

2. Vertex Insertion

   There is a vertex ID v that exists in G_t^* but not in $G*_1$. For all the vertices u which are the automorphic mappings for v from the AVT for G_t^*, the new vertex ID v is inserted in the generalized ID of u.

5.6 OTHER APPROACHES

Most research has focused on node re-identification. However, Zheleva and Getoor [2007] propose to prevent re-identification of sensitive edges in graph data. Other than deterministic algorithms in graph anonymization, Hay et al. [2007] propose to protect privacy against node re-identification by subgraph knowledge, as well as vertex requirement structural queries, by random graph perturbation. Ying and Wu [2008] address the same problem of Hay et al. [2007] and Hay et al. [2008] by edge addition/deletion and switching, and they also analyze the effect of their method by studying the spectrum of a graph.

5.7 FUTURE DIRECTIONS

Most of the existing works focus on one-time publishing over the social network data. Although Zou et al. [2009] have proposed to study multiple-time publishing over an evolving social network, the anonymization cost can be quite high. More study on this topic will be needed in the future.

In general, privacy preserving publishing for data graph is a very difficult problem. There is a lot of room for improvement on the quality of the data published by better anonymization methods.

While node re-identification has been studied, the problem of link protection has received much less attention. It is believed that links are sensitive information that should also be protected. Hence one direction for future works will be to consider such protection issues.

CHAPTER 6

Other Data Types

In addition to relational tables and graphs, there are other kinds of data such as spatial and transactional data. In this chapter, we describe privacy models and anonymization techniques over theses data types.

6.1 SPATIAL DATA

In many spatial applications, users typically issue queries like "where is the nearest restaurant?" and "where is the nearest gas station?" Figure 6.1(a) shows a spatial layout of two datasets, P and U. P contains six points of interests (POIs) such as restaurants and gas stations, namely $p_1, p_2, p_3, ..., p_6$, and U contains five mobile users, namely $u_1, u_2, u_3, ..., u_5$. Consider that a user u_1 wants to issue a query to find the nearest point $p \in P$. According to Figure 6.1(a), it is easy to determine that p_1 is the answer.

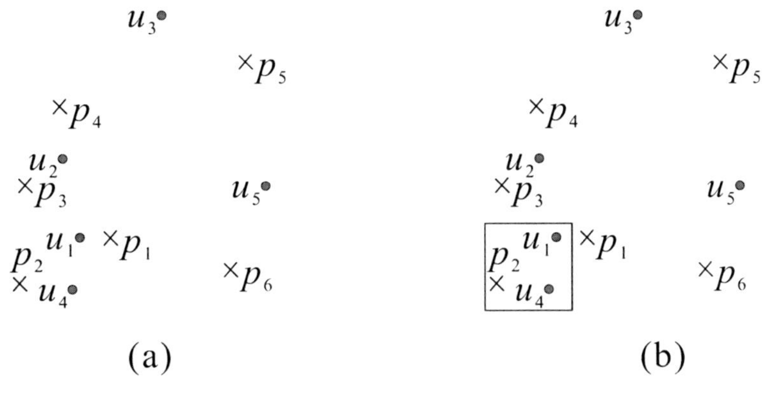

Figure 6.1: An example showing all users and all points of interests

Figure 6.2 shows how the user obtains the answer after s/he issues a query. In this figure, the location of a mobile user is detected by a base station. Then, the user can perform a query via the base station. The base station then transfers the user query to a *location-based service (LBS) server* that can process the query. After processing the query, the server passes the (exact) answer of the query to the user via the base station.

In many applications, a mobile user does not want to disclose his/her own location to the LBS server. Typically, the base station is a *trusted party*, which means that it will not pass the location of

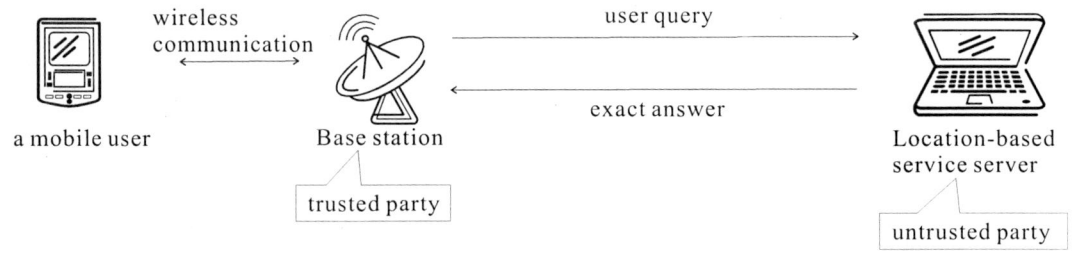

Figure 6.2: Location based service when location privacy is not protected

the users to other parties, and thus it will maintain the location privacy of users. However, the LBS server is an *untrusted party*. This means that it may disclose the location of users to public or other parties.

In the following, we describe two ways of protecting the location privacy of mobile users.

- *Approach with anonymizer:* The first way is to make use of an additional component called an *anonymizer* that plays a mediator role to communicate between the trusted base station and the untrusted LBS server. The objective of an anonymizer is to cloak the exact location of the mobile user with a larger region such that the LBS server cannot deduce the exact location. We will describe this technique in Section 6.1.1.

- *Approach without anonymizer:* The second way is to make use of cryptographic techniques such that the query with the exact location of a user is encrypted and thus the server does not know the exact query and the location. Since cryptographic techniques are secure, this approach does not involve any anonymizer. Details can be found in Section 6.1.2.

6.1.1 WITH ANONYMIZER

One system that uses an anonymizer is *Casper* [Mokbel et al., 2007]. As depicted in Figure 6.3, the base station is a trusted party, and thus a user can securely communicate with it. Here, the anonymizer is also trusted. It receives user queries (in particular, the nearest neighbor of the user in P) and generates a larger region that covers the user's location. The region is called a *cloaked spatial region*. Then, the anonymizer passes this cloaked region to the LBS server. Since the region is a large area instead of a particular location and can be regarded as uncertain, the server returns a set of possible candidate answers to the anonymizer. Since the anonymizer knows the exact location, it will filter out the answer returned from the server and return the exact answer to the user via the base station.

Let us illustrate the idea with an example. Suppose user u_1 in Figure 6.1(a) wants to find his/her nearest neighbor in P. Under this framework, the anonymizer will generate a large region like the one as shown in Figure 6.1(b) and then pass it to the LBS server. Since this query involves a region, the LBS server will return all possible points in P that are potential answers, namely p_1, p_2

6.1. SPATIAL DATA

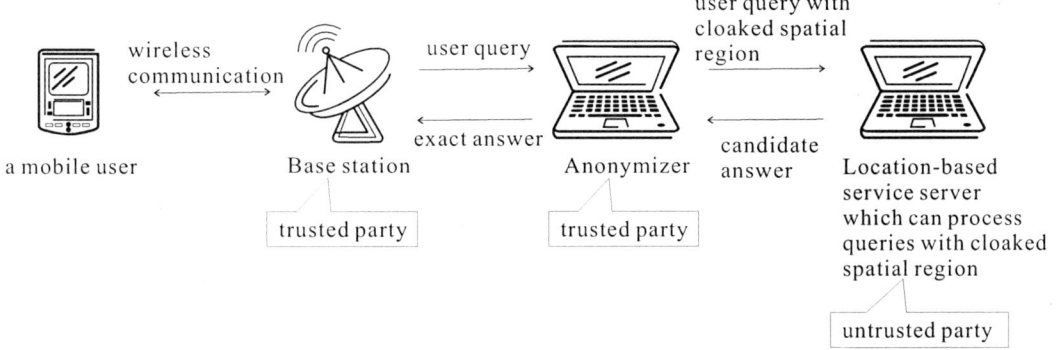

Figure 6.3: Casper system for location-based service

and p_3. Then, the anonymizer receives the set of potential answers and determines that the exact nearest neighbor is p_1.

There are two questions raised here:

- How can the cloaked region protect the location privacy of a mobile user?

- How large should the cloaked region be in order to protect the location privacy of a mobile user?

The first question can be easily answered by using the concept of k-anonymity we discussed in Chapter 2, together with the minimum region R_{min} a user needs. Specifically, the cloaked region should satisfy two requirements. The first requirement is for k-anonymity: if the cloaked region for a user u contains at least k users, we say that the cloaked region satisfies k-anonymity. The second requirement is for the minimum region a user needs: the size of the cloaked region must be at least R_{min}. Under this framework, since the cloaked region in each query satisfies k-anonymity and has size at least R_{min}, even though the adversary obtains the query region, the adversary can deduce only that each user in the region may have issued this query with probability at most $1/k$. Thus, the location privacy of a user can be protected.

The second question may be answered by generating a *minimal* region that covers the mobile user and $k - 1$ other users. It is easy to see that, if the region is the whole space, then the region contains all users. Typically, the number of users is much larger than k. Although privacy can be protected, in this case, the number of the candidate answers returned by the LBS server will be extremely large.

Instead, the objective is to find a minimal region that covers k users including the mobile user who issues the query. Casper uses a pyramid-like structure to represent the region. The whole space (Figure 6.4(a)) is partitioned into four quadrants as shown in Figure 6.4(b). Each of these quadrants are further partitioned into four sub-quadrants as shown in Figure 6.4(c). The partitioning forms a

106 6. OTHER DATA TYPES

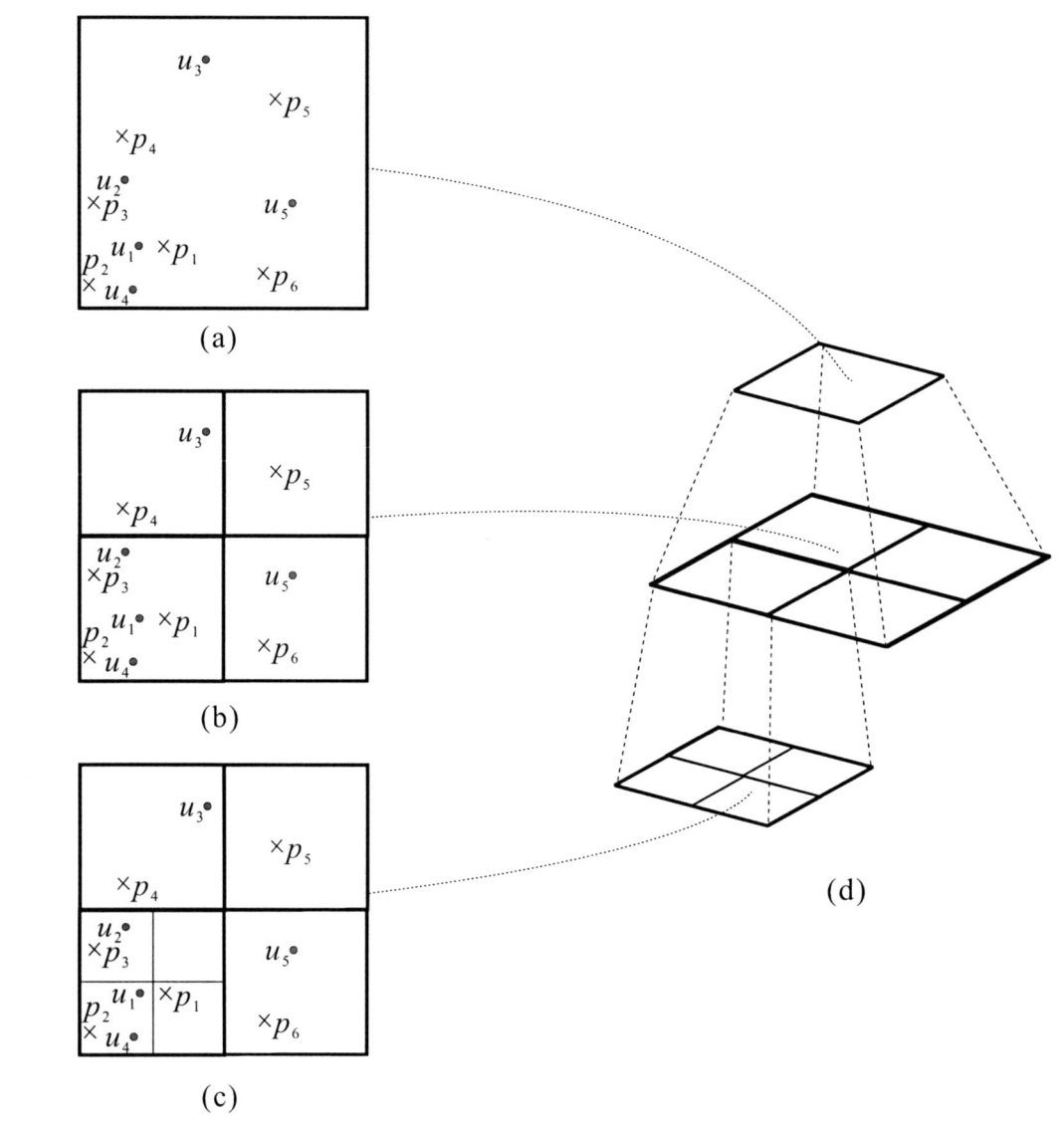

Figure 6.4: A pyramid

pyramid as shown in Figure 6.4(d). The depth of the pyramid is determined by the anonymizer and is an input parameter.

Suppose user u_1 wants to perform a query. If $k = 2$, then the anonymizer will generate the bottom-leftmost sub-quadrant (in Figure 6.4(c)) as a cloaked region.

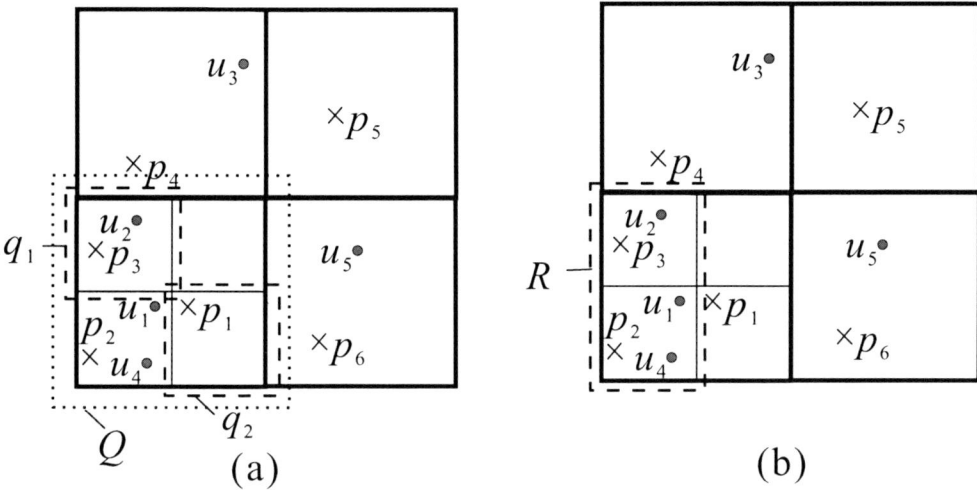

Figure 6.5: Illustration of algorithm Casper

However, in general, it is possible that the small grid does not contain at least k users. In this case, Casper will generate a larger grid as follows. In our example, there are four sub-quadrants in a quadrant Q. If $k = 3$, for example, we know that the bottom-leftmost sub-quadrant does not contain at least 3 users. Casper will find each sub-quadrant that is adjacent to the sub-quadrant containing u_1. Note that there are two adjacent sub-quadrants. One is at the top-left corner q_1 of Q and the other is at the bottom-right corner q_2 of Q. Figure 6.5(a) shows quadrant Q and the two adjacent sub-quadrants q_1 and q_2. It searches one sub-quadrant to see if there are any users; if there are, then it forms a rectangle by taking a union of the two sub-quadrants. If there are no users in this sub-quadrant, it tries to the other one. Let us assume in our example. It searches the bottom-right sub-quadrant q_2 first. Since there are no users in q_2, it searches for the users in the top-left sub-quadrant q_1. Since this sub-quadrant contains one additional user, Casper will generate a rectangle R that is formed by performing a union operation between this top-left sub-quadrant and the bottom-left sub-quadrant. This rectangle is the cloaked region that contains at least 3 users. Figure 6.5(b) shows the resulting rectangle R.

If Casper cannot find sufficient users in the adjacent sub-quadrants, it will perform this process recursively by performing the search over the quadrant Q (instead of the sub-quadrant).

6.1.2 WITHOUT ANONYMIZER

The anonymizer may become a major source of privacy attacks by an adversary [Ghinita et al., 2008]. Cryptographic techniques can be used in order to remove the anonymizer. Figure 6.6 shows the framework of this privacy model. In this case, as before, a mobile user communicates with the trusted base station securely. However, the base station sends an *encrypted* form of the user query to the LBS server. This encryption has the good feature that the LBS server does not know the original content of the user query. When the LBS server answers this encrypted user query by following the cryptographic protocol, it returns the answers to the users in an encrypted form. Note that, since the query and the answer are both encrypted, the LBS server does not know the exact query and the exact answer. Finally, after the user receives the encrypted answer, s/he can decrypt it easily and obtain the exact answer.

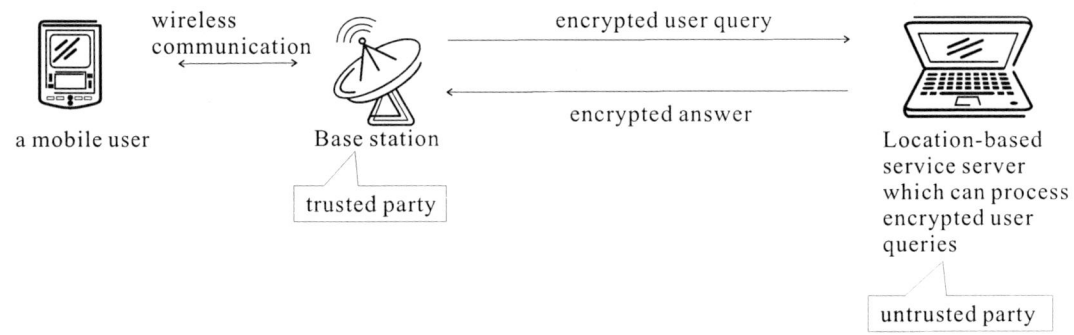

Figure 6.6: A system for location based service by using cryptographic techniques

6.2 TRANSACTIONAL DATA

In Chapter 2 and Chapter 3, we discussed how to protect individual privacy in a (relational) table. A relational table is associated with a set of quasi-identifier attributes and a sensitive attribute where each attribute is a *single* value. A transactional data studied in this chapter is also associated with a set of quasi-identifier attributes and a sensitive attribute where each attribute is a *sequence* of values or a *set* of values (instead of a single value in the context of the relational table). For example, in the relational table, an individual can have attribute Gender equal to male. But, in the transactional data, an individual can have attribute Items Purchased equal to a sequence of items such as "Diaper, Beer" (which means that the individual purchased diaper and then beer).

In the following, for the sake of illustration, we focus on a special case of transactional data where there is only one quasi-identifier attribute called *Public Values* and only one sensitive attribute called *Private Value* where the quasi-identifier attribute corresponds to a sequence of values and the sensitive attribute corresponds to a single value only.

Each tuple can be represented by a sequence of values (or items) which is stored in attribute Public Values. There are many applications that have transaction data. Some examples are search engine data where a sequence of keywords forms a transaction and online shopping data where a sequence of item purchases forms a transaction.

In order to protect privacy over transactional data, a privacy model called (h, k, p)-*coherent* is defined [Xu et al., 2008a,b] where h and p are two real numbers ranging from 0 to 1 and k is a positive integer. This model involves two kinds of values, *public values* and *private value*. Public values can be regarded as the quasi-identifier values in the context of the traditional table, and private values can be regarded as the sensitive value. Table 6.1 shows an example of transactional data where a, b, c, d and e are public values and HIV, Heart Disease, Cancer and Diabetes are private values.

Table 6.1: Transactional data

Name	Public values	Private value
Peter	a, b	HIV
Mary	a, c	HIV
Chris	b	Heart Disease
David	d, b	Diabetes
Emily	d, e	Cancer

Table 6.2: Transactional data satisfying $(0.5, 2, 2)$-coherence

Name	Public values	Private value
Peter	$*, b$	HIV
Mary	$*, *$	HIV
Chris	b	Heart Disease
David	$*, b$	Diabetes
Emily	$*, *$	Cancer

Since k-anonymity and l-diversity are commonly adopted in the context of relational tables, the privacy model, (h, k, p)-coherence, is proposed to give the privacy protection provided by both the k-anonymity requirement and the l-diversity requirement. Roughly speaking, a table is said to satisfy (h, k, p)-coherence if the table satisfies the k-anonymity requirement and the l-diversity requirement in the context of transaction data.

- *k-anonymity:* Under this model, an adversary can uniquely identify an individual that has at most p public values where p is a positive integer (given by the data owner for privacy protection). For example, if p is 2, the adversary has the knowledge that Peter has two public values, namely a and b, s/he can uniquely identify the tuple for Peter by his two public values in Table 6.1, because there is only one tuple in Table 6.1 containing a and b. Similarly, Mary can be uniquely identified by only one public value c. We define the *identifying set* to be a set containing at most p public values. This identification problem with the identifying set can be solved by using the concept of k-anonymity. Specifically, the model ensures that, for each possible identifying set X, there exist at least k tuples containing X.

- *l-diversity:* Even though the identification problem can be solved by using the k-anonymity concept, there is no protection of private values. If $p = 1$, the adversary knows that Peter has a public value a, and Peter's tuple may be either the first tuple or the second tuple. However, the adversary is sure that Peter suffers from HIV because both tuples are linked to HIV. In order to avoid this sensitive inference, l-diversity is adopted. Specifically, their model guarantees that the probability that each possible identifying set is linked to any private value is at most h where h is a given parameter.

Formally, a table is said to satisfy (h, k, p)-*coherence* (or the table is (h, k, p)-*coherent*) if the table satisfies the following:

- For each possible identifying set X, there exist at least k tuples containing X.

- The probability that each possible identifying set is linked to any private value is at most h where h is a given parameter.

After these two privacy requirements are adopted, if $h = 0.5$, $k = 2$ and $p = 2$, the data owner will publish the table as shown in Table 6.2 that satisfies $(0.5, 2, 2)$-coherence.

A three-step algorithm is proposed to generate a (h, k, p)-coherent table [Xu et al., 2008a].

- **Step 1:** The algorithm finds the set X of all identifying sets that violate one of the above two conditions.

- **Step 2:** Using some heuristics functions (whose details are not important), it chooses an identifying set S from X and then an item from S. If the item chosen is not equal to *, the algorithm suppresses this item to * in the data. (NOTE: Symbol * corresponds a fully generalized value.) Otherwise, it randomly chooses an item that is not equal to * and suppresses this item to * in the data.

- **Step 3:** After the suppression operation, it updates set X. If X is non-empty, Step 2 is performed again. Otherwise, the algorithm terminates. The resulting suppressed table satisfies (h, k, p)-coherence.

Let us illustrate this algorithm with Table 6.1 to generate Table 6.2 satisfying $(0.5, 2, 2)$-coherence. Consider Step 1. Initially, we find a set A of all identifying sets that violate the k-anonymity requirement where $k = 2$. It is easy to verify that $A = \{\{c\}, \{e\}, \{a, b\}, \{a, c\}, \{d, b\}, \{d, e\}\}$. Besides, we find a set B of all identifying sets that violate the l-diversity requirement where $l = 1/0.5 = 2$. It is easy to verify that $B = \{\{a\}, \{c\}, \{e\}, \{a, b\}, \{a, c\}, \{d, b\}, \{d, e\}\}$. X is equal to the set of all identifying sets that violate the k-anonytmity requirement or the l-diversity requirement. Thus, $X = \{\{a\}, \{c\}, \{e\}, \{a, b\}, \{a, c\}, \{d, b\}, \{d, e\}\}$.

Consider Step 2. According to some heuristics functions, we select $\{a\}$ from X and suppress it to *. Table 6.1 becomes Table 6.3(a). X is now updated to $X = \{\{*\}, \{c\}, \{e\}, \{*, b\}, \{*, c\}, \{d, b\}, \{d, e\}\}$.

Since X is non-empty, we still need to continue to re-execute Step 2. According to some heuristics functions, we select c and suppress it to *. Similarly, X is non-empty, we need to execute

Table 6.3: Illustration for the algorithm generating (0.5, 2, 2)-coherent transactional data

Name	Public values	Private value
Peter	$*, b$	HIV
Mary	$*, c$	HIV
Chris	b	Heart Disease
David	d, b	Diabetes
Emily	d, e	Cancer

(a) After suppressing a

Name	Public values	Private value
Peter	$*, b$	HIV
Mary	$*, *$	HIV
Chris	b	Heart Disease
David	d, b	Diabetes
Emily	$d, *$	Cancer

(b) After suppressing c and e

Name	Public values	Private value
Peter	$*, b$	HIV
Mary	$*, *$	HIV
Chris	b	Heart Disease
David	$*, b$	Diabetes
Emily	$*, *$	Cancer

(c) After suppressing d

Step 2 again. This time, we select e and suppress it to $*$. Table 6.3(b) becomes Table 6.3(c). X becomes $\{\{*\}, \{e\}, \{*, b\}, \{*, *\}, \{d, b\}, \{d, *\}\}$.

X is still non-empty. We select d and suppress it to $*$ as shown in Table 6.3(d). Now, X is equal to $\{\}$. Since X is an empty-set, we can terminate the process and the resulting table satisfies (0.5, 2, 2)-coherence.

6.3 CONCLUSION

In this chapter, we learnt that data publishing is applicable to not only relational tables but also other forms of data. Some examples illustrated in this chapter are spatial data (Section 6.1) and transactional data (Section 6.2).

Although these forms of data are different from relational tables, the *principle* of protecting individual privacy for relational tables is still applicable to these forms. For example, the principle of k-anonymity is still adopted in spatial data and transactional data although the real detailed implementation is a little bit different.

CHAPTER 7
Future Research Directions

Privacy-preserving data publishing is a promising research area since individual privacy has become a major concern. In this book, we first discussed a variety of anonymization techniques over traditional relational tables. Then, we described how the data owner anonymizes the data when the data is published multiple times. Finally, we discussed the privacy models over the data of other types.

There are a lot of interesting research directions. We describe some possible future directions in this chapter.

7.1 ONE-TIME DATA PUBLISHING

Although one-time data publishing has been studied for some time, there are still a lot of open problems. The first open problem is to study how a mixture of different kinds of background knowledge affects the privacy model. In Chapters 2 and 3, we considered the following types of background knowledge *independently*. We know that the adversary can breach individual privacy with one kind of the background knowledge:

- the knowledge about the quasi-identifier of individuals,
- the knowledge about the distribution of sensitive values,
- the knowledge about the linkage of individuals to sensitive values,
- the knowledge about the relationship among individuals,
- the knowledge about anonymization,
- the knowledge mined from the microdata, and
- the knowledge mined from the published data.

In reality, an adversary can have more than one kind of background knowledge. It is interesting to design a complete model to avoid any privacy breaches in this case.

The second open problem is to study other kinds of background knowledge that have not been studied in the literature. For instance, one possible background knowledge is that a sensitive value like lung cancer may be similar to another sensitive value like breast cancer. Since lung cancer and breast cancer are related to cancer, if a QI-group contains two individuals and these two sensitive values, the adversary can figure out that each individual is linked to cancer. It is interesting to study how to protect individual privacy in the presence of multiple "similar" sensitive values.

The third open problem is to improve the efficiency of the anonymization process. In Chapters 2 and 3, we only focused on describing how the modified data can protect individual privacy. However, how to modify the data *efficiently* is a promising direction since there are not much studies on this direction.

The fourth open problem is to investigate how to analyze the anonymized data. In Section 3.8, we described some methods to analyze the anonymized data. In fact, these methods are quite preliminary. There is still room to design other analysis methods. For example, if the table is anonymzied with global recoding, a classifier can be generated based on this table because all values of each attribute have the same domain (e.g., all values of attribute Zipcode have the format DDD** where D is an exact digit and * is a special symbol denoting a generalized digit.) However, if the table is anonymized with local recoding, it is not straightforward how to generate a classifier based on this table because different values of the same attribute may come from different domains (e.g., in attribute Zipcode, there exists a value like 543** and another value like 5432*).

7.2 MULTIPLE-TIME DATA PUBLISHING

Multiple-time data publishing is a new topic. There are a lot of issues that have not been studied.

Firstly, how the background knowledge discussed in the context of one-time data publishing can be used in multiple-time data publishing for privacy protection is a possible direction. Some examples are the knowledge about the linkage of individuals to sensitive values and the knowledge about the relationship among individuals. These kinds of knowledge have not been studied in the context of multiple-time data publishing.

Secondly, in Section 4.2, we described how the data publisher protects individual privacy by considering the sensitive value-based correlation. We studied two kinds of sensitive values for this correlation, namely permanent sensitive values and transient sensitive values. If an individual is linked to a permanent sensitive value in a previous release, s/he is linked to this sensitive value in a later release with 100% probability. If an individual is linked to a transient sensitive value in a previous release, s/he may or may not be linked to this sensitive value in a later release. In some cases, some sensitive values linked to individuals remain unchanged with a certain probability. For example, an individual who suffered from lung cancer can be cured after 1 year with 50% probability. We call this probability associated with lung cancer the *transient probability*. If the transient probability associated with a sensitive value is equal to 100%, this sensitive value becomes the transient sensitive value we studied. If it is equal to 0%, it becomes the permanent sensitive value. If this probability is larger, it is more likely that the sensitive value changes freely. It is interesting to design a privacy model by considering the transient probability as a possible direction.

7.3 PUBLISHING GRAPH DATA

Most of the existing works about publishing social network (or graph data) focus on one-time publishing over the social network data. Although Zou et al. [2009] have proposed to study multiple-time publishing over an evolving social network, the anonymization cost can be quite high. More work is needed on this topic.

In general, privacy preserving publishing for data graph is a very difficult problem. There is a lot of room for improvement on the quality of the data published by better anonymization methods.

While node re-identification has been studied, the problem of link protection has received much less attention. It is believed that links are sensitive information that should also be protected. Hence one direction for future works will be to consider such protection issues.

7.4 PUBLISHING DATA OF OTHER FORMS

When we consider the privacy issues over data of other types, the anonymization techniques and the privacy models discussed in the traditional relational model may not be adopted *directly*. For example, k-anonymity and l-diversity originally designed for relational tables have been adopted in the data of other types such as spatial data and transactional data. There are some other popular privacy models such as t-closeness that have not been applied to these data types. It is interesting to study how we can implement these privacy models in the data of other types.

In addition, since data change over time, it is interesting to study how the anonymization techniques originally designed for the relational data can be applied on other data types. Since these data may have some properties that do not exist in relational tables, there are some interesting issues. For example, in relational tables, an individual can change his/her occupation *freely* over time. However, in spatial data, an object (or an individual) cannot move freely over time. Specifically, it can move at most a certain distance from its original location. With this bounded movement, we can anonymize the spatial data with less information loss.

APPENDIX A

Definition of Entropy l-Diversity and Recursive l-Diversity

In this appendix, we define entropy l-diversity, and recursive l-diversity (more specifically, recursive (c, l)-diversity).

Entropy l-diversity

Let us first describe entropy l-diversity. Let S be a set of values in the sensitive attribute. Let $|S|$ be the total number of possible values in S. Consider an QI-group G. Let p_s be the frequency (in fraction) of the value s in group G where $s \in S$. The entropy of a QI-group G is defined to be

$$-\sum_{s \in S} p_s Prob(p_s)$$

Intuitively, if the entropy of a QI-group is larger, then it is more likely that the frequency of a value in S is similar to that of another value in S. Thus, it is more difficult for the adversary to figure out the frequency of a value in S. This motivates the definition of entropy l-diversity as follows.

Definition A.1 Entropy l-diversity. A QI-group G is said to satisfy *entropy l-diversity* if the entropy of G is at least $\log l$. A table is said to satisfy *entropy l-diversity* if each QI-group in the table satisfies entropy l-diversity.

However, Machanavajjhala et al. [2006] suggest that since $-x \log x$ is a concave function, if the table satisfies *entropy l-diversity*, then the entropy of the QI-group containing all tuples in the table must be at least $\log l$. This privacy is too restrictive. This is because this requirement is hardly achieved since in real life applications, usually, some values like flu in the sensitive attribute are common and may appear more than 90% of tuple, which means that the entropy of the QI-group containing all tuples in the table is usually smaller than $\log l$.

Recursive l-diversity (or (c, l)-diversity)

Due to the weakness of the entropy l-diversity, Machanavajjhala et al. [2006] propose a privacy model called (c, l)-*diversity* where c is a non-negative real number and l is a positive integer.

Consider a QI-group G. Let n_i be the frequency (in count) of the i-th most frequent value in group G.

Definition A.2 Recursive (c,l)-diversity. An QI-group G is said to satisfy *recursive (c,l)-diversity* if $n_1 < c(n_l + n_{l+1} + ... + n_{|S|})$. A table is said to satisfy *recursive (c,l)-diversity* if each QI-group in the table satisfies recursive (c,l)-diversity.

The intuition of recursive (c,l)-diversity is that, after the adversary removes $l-1$ possible values of S in a QI-group G, s/he cannot infer any sensitive value of an individual.

Bibliography

C. C. Aggarwal and P. S. Yu. A survey of uncertain data algorithms and applications. *IEEE Trans. Knowl. and Data Eng.*, 21(5):609–623, 2009. DOI: 10.1109/TKDE.2008.190 66

C. C. Aggarwal and P. S. Yu. A condensation approach to privacy preserving data mining. In *Advances in Database Technology, Proc. 9th Int. Conf. on Extending Database Technology*, pages 183–199, 2004. DOI: 10.1007/b95855 17

G. Aggarwal, T. Feder, K. Kenthapadi, R. Motwani, R. Panigrahy, D. Thomas, and A. Zhu. Anonymizing tables. In *Proc. 10th Int. Conf. on Database Theory*, pages 246–258, 2005. DOI: 10.1007/b104421 19

R. Agrawal and R. Srikant. Fast algorithms for mining association rules. In *Proc. 20th Int. Conf. on Very Large Data Bases*, pages 487–499, 1994. 55

L. Backstrom, C. Dwork, and J. Kleinberg. Wherefore art thou r3579x? anonymized social networks, hidden patterns and structural steganography. In *Proc. 16th Int. World Wide Web Conf.*, Banff, Alberta, May 2007. 92

Lars Backstrom, Daniel P. Huttenlocher, Jon M. Kleinberg, and Xiangyang Lan. Group formation in large social networks: membership, growth, and evolution. In *Proc. 12th ACM SIGKDD Int. Conf. on Knowledge Discovery and Data Mining*, pages 44–54, 2006. DOI: 10.1145/1150402.1150412 93

M. Barbaro and T. Z. Jr. A face is exposed for AOL searcher no. 4417749. In *New York Times*, 2006. 3

R. Bayardo and R. Agrawal. Data privacy through optimal k-anonymization. In *Proc. 21st Int. Conf. on Data Engineering*, pages 217–228, 2005. DOI: 10.1109/ICDE.2005.42 64

E. Beinat. Privacy and location-based: Stating the policies clearly. In *Proc. of the International Conference on GeoInformatics*, pages 14–17, 2001. 4

B. Bollobas. *Random Graphs*. Cambridge, 2001. 92

Y. Bu, A. Fu, R. Wong, L. Chen, and J. Li. Privacy preserving serial data publishing by role composition. In *Proc. 34th Int. Conf. on Very Large Data Bases*, pages 845–856, 2008. DOI: 10.1145/1453856.1453948 83

D. Burdick, P.M. Deshpande, T.S. Jayram, R. Ramakrishnan, and S. Vaithyanathan. OLAP over uncertain and imprecise data. In *Proc. 31st Int. Conf. on Very Large Data Bases*, pages 123–144, 2005. DOI: 10.1007/s00778-006-0033-y 59

D. Burdick, A. Doan, R. Ramakrishnan, and S. Vaithyanathan. OLAP over imprecise data with domain constraints. In *Proc. 33rd Int. Conf. on Very Large Data Bases*, pages 39–50, 2007. 59

J. Byun, Y. Sohn, E. Bertino, and N. Li. Secure anonymization for incremental datasets. In *Secure Data Management*, pages 48–63, 2006. DOI: 10.1007/11844662_4 83

Alina Campan and Traian Marius Truta. A clustering approach for data and structural anonymity in social networks. In *2nd ACM SIGKDD International Workshop on Privacy, Security and Trust in KDD*, 2008. DOI: 10.1007/978-3-642-01718-6_4 97

D. M. Carlisle, M. L. Rodrian, and C. L. Diamond. California inpatient data reporting manual medical information reporting for California. Technical report, 2007. 1

B.-C. Chen, K. LeFevre, and R. Ramakrishnan. Privacy skyline: Privacy with multidimensional adversarial knowledge. In *Proc. 33rd Int. Conf. on Very Large Data Bases*, pages 770–781, 2007. 30, 44, 45, 46, 47

James Cheng, Yiping Ke, Wilfred Ng, and Jeffrey Xu Yu. Context-aware object connection discovery in large graphs. *Proc. 25th Int. Conf. on Data Engineering*, 2009. DOI: 10.1109/ICDE.2009.87 93

G. Cormode, F. Li, and K. Yi. Semantics of ranking queries for probabilistic data and expected ranks. In *Proc. 25th Int. Conf. on Data Engineering*, pages 305–316, 2009. DOI: 10.1109/ICDE.2009.75 66

P. Erdos and T. Gallai. Graphs with prescribed degrees of vertices. *Mat. Lapok*, 11:264–274, 1960. 95

Christos Faloutsos, Kevin S. McCurley, and Andrew Tomkins. Fast discovery of connection subgraphs. In *Proc. 10th ACM SIGKDD Int. Conf. on Knowledge Discovery and Data Mining*, pages 118–127, 2004. DOI: 10.1145/1014052.1014068 93

B. C. M. Fung, K. Wang, and P. S. Yu. Top-down specialization for information and privacy preservation. In *Proc. 21st Int. Conf. on Data Engineering*, pages 205–216, 2005. DOI: 10.1109/ICDE.2005.143 33, 64

G. Ghinita, P. Kalnis, A. Khoshgozaran, C. Shahabi, and K.-L. Tan. Private queries in location based services: Anonymizers are not necessary. In *Proc. ACM SIGMOD Int. Conf. on Management of Data*, pages 121–132, 2008. DOI: 10.1145/1376616.1376631 107

M. Hay, G. Miklau, D. Jensen, P. Weis, and S. Srivastava. Anonymizing social networks. Technical report, University of Massachusetts Amherst, 2007. 90, 101

M. Hay, G. Miklau, D. Jensen, D. Towsley, and P. Weis. Resisting structural re-identification in anonymized social networks. In *Proc. 34th Int. Conf. on Very Large Data Bases*, pages 102–114, 2008. DOI: 10.1145/1453856.1453873 91, 92, 93, 96, 97, 98, 101

M. Hua, J. Pei, W. Zhang, and X. Lin. Ranking queries on uncertain data: A probabilistic threshold approach. In *Proc. ACM SIGMOD Int. Conf. on Management of Data*, pages 673–686, 2008. DOI: 10.1145/1559845.1559861 66

D. Kifer. Attacks on privacy and deFinetti's theorem. In *Proc. ACM SIGMOD Int. Conf. on Management of Data*, pages 127–138, 2009. 30, 58, 59

D. Kifer and J. Gehrke. Injecting utility into anonymized datasets. In *Proc. ACM SIGMOD Int. Conf. on Management of Data*, pages 217–228, 2006. DOI: 10.1145/1142473.1142499 27

J. Kim and W. Winkler. Preserving individual privacy in serial data publishing. In *the ASA Section on Survey Research Methods*, pages 114–119, 1995. 15

Gueorgi Kossinets, Jon M. Kleinberg, and Duncan J. Watts. The structure of information pathways in a social communication network. In *Proc. 14th ACM SIGKDD Int. Conf. on Knowledge Discovery and Data Mining*, pages 435–443, 2008. DOI: 10.1145/1401890.1401945 93

Ravi Kumar, Jasmine Novak, and Andrew Tomkins. Structure and evolution of online social networks. In *Proc. 12th ACM SIGKDD Int. Conf. on Knowledge Discovery and Data Mining*, pages 611–617, 2006. DOI: 10.1145/1150402.1150476 93

M. Kuramochi and G. Karypis. Finding frequent patterns in a large sparse graph. *Data Mining and Knowledge Discovery*, 11(3):243–271, 2005. DOI: 10.1007/s10618-005-0003-9 99

K. LeFevre, D. J. DeWitt, and R. Ramakrishnan. Incognito: Efficient full-domain k-anonymity. In *Proc. ACM SIGMOD Int. Conf. on Management of Data*, pages 49–60, 2005. DOI: 10.1145/1066157.1066164 19, 29, 30, 47, 48, 86

K. LeFevre, D. DeWitt, and R. Ramakrishnan. Mondrian multidimensional k-anonymity. In *Proc. 22nd Int. Conf. on Data Engineering*, page 25, 2006. DOI: 10.1109/ICDE.2006.101 33, 36, 48, 64, 86

K. Lefevre, D. J. Dewitt, and R. Ramakrishnan. Workload-aware anonymization techniques for large-scale datasets. *ACM Trans. Database Syst.*, 33(3):1–47, 2008. DOI: 10.1145/1386118.1386123 64

Jure Leskovec, Lars Backstrom, Ravi Kumar, and Andrew Tomkins. Microscopic evolution of social networks. In *Proc. 14th ACM SIGKDD Int. Conf. on Knowledge Discovery and Data Mining*, pages 462–470, 2008. DOI: 10.1145/1401890.1401948 93

J. Li, Y. Tao, and X. Xiao. Preservation of proximity privacy in publishing numerical sensitive data. In *Proc. ACM SIGMOD Int. Conf. on Management of Data*, pages 473–486, 2008. DOI: 10.1145/1376616.1376666 23, 30

N. Li and T. Li. t-closeness: Privacy beyond k-anonymity and l-diversity. In *Proc. 23rd Int. Conf. on Data Engineering*, pages 106–115, 2007. DOI: 10.1109/ICDE.2007.367856 19, 30, 40, 41, 59

T. Li and N. Li. Injector: Mining background knowledge for data anonymization. In *Proc. 24th Int. Conf. on Data Engineering*, pages 446–455, 2008. DOI: 10.1109/ICDE.2008.4497453 30, 40, 44, 52, 54, 61

T. Li, N. Li, and J. Zhang. Modeling and integrating background knowledge in data anonymization. In *Proc. 25th Int. Conf. on Data Engineering*, pages 6–17, 2009. DOI: 10.1109/ICDE.2009.86 30, 40, 43, 44

K. Liu and E. Terzi. Towards identity anonymization on graphs. In *Proc. ACM SIGMOD Int. Conf. on Management of Data*, pages 93–106, 2008. DOI: 10.1145/1376616.1376629 94

A. Machanavajjhala, J. Gehrke, and D. Kifer. l-diversity: privacy beyond k-anonymity. In *Proc. 22nd Int. Conf. on Data Engineering*, page 24, 2006. 19, 20, 30, 40, 44, 47, 59, 64, 117

D. J. Martin, D. Kifer, A. Machanavajjhala, and J. Gehrke. Worst-case background knowledge for privacy-preserving data publishing. In *Proc. 23rd Int. Conf. on Data Engineering*, pages 126–135, 2007. DOI: 10.1109/ICDE.2007.367858 44, 46

A. McCallum, A. Corrada-Emmanuel, and X. Wang. Topic and role discovery in social networks. In *Proc. 19th Int. Joint Conf. on AI*, 2005. 93

M. F. Mokbel, C.-Y. Chow, and W. G. Aref. The new Casper: Query processing for location services without compromising privacy. In *Proc. 33rd Int. Conf. on Very Large Data Bases*, pages 763–774, 2007. 104

A. Narayanan and V. Shmatikov. How to break anonymity of the Netflix prize data set. In *ArXix*, 2006. URL http://arxiv.org/abs/cs/0610105. 3

M. Nergiz, M. Atzori, and C.W. Clifton. Hiding the presence of individuals from shared databases. In *Proc. ACM SIGMOD Int. Conf. on Management of Data*, pages 665–676, 2007. DOI: 10.1145/1247480.1247554 83

J. Pei, Y. Tao, J. Li, and X. Xiao. Privacy preserving publishing on multiple quasi-identifiers. In *Proc. 25th Int. Conf. on Data Engineering*, pages 127–138, 2009. DOI: 10.1109/ICDE.2009.183 25, 30

S. P. Reiss. Practical data-swapping: the first steps. *ACM Trans. Database Syst.*, 9(1):20–37, 1984. DOI: 10.1145/348.349 16

S. P. Reiss, M. J. Post, and T. Dalenius. Non-reversible privacy transformations. In *Proc. 1st ACM SIGACT-SIGMOD Symp. on Principles of Database Systems*, pages 139–146, New York, NY, USA, 1982. DOI: 10.1145/588111.588134 16

P. Samarati. Protecting respondents' identities in microdata release. *IEEE Trans. Knowl. and Data Eng.*, 13(6):1010–1027, 2001. DOI: 10.1109/69.971193 29, 30

P. Samarati and L. Sweeney. Protecting privacy when disclosing information: k-anonymity and its enforcement through generalization and suppression, unpublished manuscript. 1998. URL http://citeseer.ist.psu.edu/samarati98protecting.html. 8

M. A. Soliman, I. F. Ilyas, and K. C.-C. Chang. Top-k query processing in uncertain databases. In *Proc. 23rd Int. Conf. on Data Engineering*, pages 896–905, 2007. DOI: 10.1109/ICDE.2007.367935 66

L. Sweeney. Achieving k-anonymity privacy protection using generalization and suppression. *International Journal on Uncertainty, Fuzziness and knowldege based systems*, 10(5):571–588, 2002a. 9

L. Sweeney. k-anonymity: a model for protecting privacy. *International Journal on Uncertainty, Fuzziness and knowldege based systems*, 10(5):557–570, 2002b. DOI: 10.1142/S0218488502001648 3, 8, 19, 30, 48, 86

Y. Tao, X. Xiao, J. Li, and D. Zhang. On anti-corruption privacy preserving publication. In *Proc. 24th Int. Conf. on Data Engineering*, pages 725–734, 2008. DOI: 10.1109/ICDE.2008.4497481 30, 45, 46

Hanghang Tong, Christos Faloutsos, Brian Gallagher, and Tina Eliassi-Rad. Fast best-effort pattern matching in large attributed graphs. In *Proc. 13th ACM SIGKDD Int. Conf. on Knowledge Discovery and Data Mining*, pages 737–746, 2007. DOI: 10.1145/1281192.1281271 93

K. Wang and B. Fung. Anonymizing sequential releases. In *Proc. 12th ACM SIGKDD Int. Conf. on Knowledge Discovery and Data Mining*, pages 414–423, 2006. DOI: 10.1145/1150402.1150449 48

K. Wang and B. C. M. Fung. Anonymizing sequential releases. In *Advances in Database Technology, Proc. 11th Int. Conf. on Extending Database Technology*, pages 414–423, 2008. 69, 71

K. Wang, P. S. Yu, and S. Chakraborty. Bottom-up generalization: A data mining solution to privacy protection. In *Proc. 2004 IEEE Int. Conf. on Data Mining*, pages 249–256, 2004. 29

K. Wang, B. C. M. Fung, and P. S. Yu. Handicapping attacker's confidence: An alternative to k-anonymization. *Knowledge and Information Systems: An International Journal*, 11(3):345–368, 2007. DOI: 10.1007/s10115-006-0035-5 47

K. Wang, Y. Xu, A. Fu, and R. Wong. FF-anonymity: When quasi-identifiers are missing. In *Proc. 25th Int. Conf. on Data Engineering*, pages 1136–1139, 2009. DOI: 10.1109/ICDE.2009.184 26

R. Wong, J. Li, A. Fu, and K. Wang. (alpha, k)-anonymity: An enhanced k-anonymity model for privacy-preserving data publishing. In *Proc. 12th ACM SIGKDD Int. Conf. on Knowledge Discovery and Data Mining*, pages 754–759, 2006. DOI: 10.1145/1150402.1150499 19, 20, 30, 47

R. Wong, A. Fu, K. Wang, and J. Pei. Minimality attack in privacy preserving data publishing. In *Proc. 33rd Int. Conf. on Very Large Data Bases*, pages 543–554, 2007. 59, 61, 63

R. Wong, A. Fu, K. Wang, and J. Pei. Anonymization-based attacks in privacy preserving data publishing. *ACM Trans. Database Syst.*, 34(2):1–46, 2009a. DOI: 10.1145/1538909.1538910 30, 51

R. Wong, A. Fu, K. Wang, Y. Xu, and P. S. Yu. Can the utility of anonymized data be used for privacy breaches? In *CoRR abs/0905.1755*, 2009b. URL http://arxiv.org/abs/0905.1755. 30, 58, 59

R. Wong, A. Fu, J. Liu, K. Wang, and Y. Xu. Global privacy guarantee in serial data publishing. In *Proc. 26th Int. Conf. on Data Engineering*, 2010. 85, 87

X. Xiao and Y. Tao. Personalized privacy preservation. In *Proc. ACM SIGMOD Int. Conf. on Management of Data*, pages 229–240, 2006a. DOI: 10.1145/1142473.1142500 24, 30, 48, 63, 86

X. Xiao and Y. Tao. Anatomy: Simple and effective privacy preservation. In *Proc. 32nd Int. Conf. on Very Large Data Bases*, pages 139–150, 2006b. 9, 59, 61, 62

X. Xiao and Y. Tao. m-invariance: Towards privacy preserving re-publication of dynamic datasets. In *Proc. ACM SIGMOD Int. Conf. on Management of Data*, pages 689–700, 2007. 74, 75, 82, 83

Y. Xu, B. C. M. Fung, K. Wang, A. Fu, and J. Pei. Publishing sensitive transactions for itemset utility. In *Proc. 2008 IEEE Int. Conf. on Data Mining*, pages 1109–1114, 2008a. DOI: 10.1109/ICDM.2008.98 108, 110

Y. Xu, K. Wang, A. Fu, and P.S. Yu. Anonymizing transaction databases for publication. In *Proc. 14th ACM SIGKDD Int. Conf. on Knowledge Discovery and Data Mining*, pages 767–775, 2008b. DOI: 10.1145/1401890.1401982 108

X. Yan and J. Han. Gspan: Graph-based substructure pattern mining. In *Proc. 2002 IEEE Int. Conf. on Data Mining*, 2002. 96

X. Ying and X. Wu. Randomizing social networks: a spectrum preserving approach. In *Proc. of SIAM International Conference on Data Mining*, Atlanta, GA, April 2008. 101

Q. Zhang, N. Koudas, D. Srivastava, and T. Yu. Aggregate query answering on anonymized tables. In *Proc. 23rd Int. Conf. on Data Engineering*, pages 116–125, 2007. DOI: 10.1109/ICDE.2007.367857 22, 30, 59

E. Zheleva and L. Getoor. Preserving the privacy of sensitive relationships in graph data. In *1st ACM SIGKDD International Workshop on Privacy, Security and Trust in KDD*, 2007. DOI: 10.1007/978-3-540-78478-4_9 101

B. Zhou and J. Pei. Preserving privacy in social networks against neighborhood attacks. In *Proc. 24th Int. Conf. on Data Engineering*, pages 506–515, 2008. DOI: 10.1109/ICDE.2008.4497459 89, 91, 93, 95, 96

L. Zou, L. Chen, and M. T. Özsu. K-automorphism: A general framework for privacy preserving network publication. In *Proc. 35th Int. Conf. on Very Large Data Bases*, pages 946–957, 2009. 92, 94, 98, 100, 101, 114

Authors' Biographies

RAYMOND CHI-WING WONG

Raymond Chi-Wing Wong received the B.Sc., M.Phil., and Ph.D. degrees in Computer Science and Engineering in the Chinese University of Hong Kong (CUHK) in 2002, 2004 and 2008, respectively. He joined Computer Science and Engineering of the Hong Kong University of Science and Technology as an Assistant Professor in 2008. During 2004-2005, he worked as a research and development assistant under an R&D project funded by ITF and a local company called Lifewood. He has published over 35 papers at major journals and conferences such as TODS, VLDBJ, TKDE, SIGKDD, VLDB, ICDE and ICDM. He has received over 20 awards. He served on the program committees of VLDB, CIKM, DASFAA, SDM and APWeb-WAIM, and refereed for a variety of journals. His research interests include database, data mining and security.

ADA WAI-CHEE FU

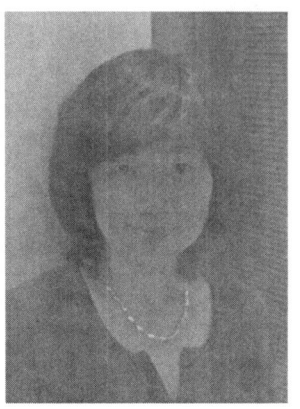

Ada Wai-Chee Fu received her B.Sc degree in computer science in the Chinese University of Hong Kong in 1983, and both M.Sc. and Ph.D. degrees in Computer Science in Simon Fraser University of Canada in 1986, 1990, respectively. She worked at Bell Northern Research in Ottawa, Canada from 1989 to 1993 on a wide-area distributed database projects. She joined the Chinese University of Hong Kong in 1993. Her research interests include database systems and data mining.

Breinigsville, PA USA
31 August 2010
244566BV00003B/2/P